Political Philosophy and Public Purpose

Series Editor

Michael J. Thompson
William Patterson University, USA

Aim of the Series

This series offers books that seek to explore new perspectives in social and political criticism. Seeing contemporary academic political theory and philosophy as largely dominated by hyper-academic and overly-technical debates, the books in this series seek to connect the politically engaged traditions of philosophical thought with contemporary social and political life. The idea of philosophy emphasized here is not as an aloof enterprise, but rather a publically-oriented activity that emphasizes rational reflection as well as informed praxis.

More information about this series at
http://www.springer.com/series/14542

Paola Cavalieri
Editor

Philosophy and the Politics of Animal Liberation

Editor
Paola Cavalieri
Milano, Italy

Philosophy and the Politics of Animal Liberation
ISBN 978-1-137-52119-4   ISBN 978-1-137-52120-0   (eBook)
DOI 10.1057/978-1-137-52120-0

Library of Congress Control Number: 2016950522

Cover image icon created by Abby Milberg from the Noun Project

Printed on acid-free paper

This Palgrave Macmillan imprint is published by Springer Nature
The registered company is Nature America Inc. New York

SERIES EDITOR'S FOREWORD

Hegel famously wrote that "philosophy is its own age apprehended in thought." He meant that philosophical reflection is forever bound to comprehending and rationally reconstructing that which already exists. Of course, Hegel's point was that this was to include what members of any historical era were conscious of and that of which they were simultaneously unaware: for it was the role of philosophical reflection to bring to light both the developed, conscious forms of life and the as yet developed, inchoate forms that existed *in potentia*. If we accept Hegel's dictum, philosophy becomes an activity that uncovers new forms of life already embryonic within the present. The notion that new values, an enlarged horizon of moral consciousness, are always being built and evolving means that we need to look at philosophy as a form of *practice*, as a kind of bringing into reality of that which is taking shape in our collective conscious awareness.

In this sense, the limits of our own moral-cultural terrain become stark once we consider the moral status of nonhuman animals. Once we see the extent to which our "evolved" moral standards and values are called into question by the ways that we dominate, abuse, exploit, and systematically destroy the nonhuman world, our ethical theories look like little more than fragile constructs, rife with contradiction. Once this is grasped, we can see that philosophy can make concrete what we have become aware of only liminally. In this sense, the relation between philosophy and politics can be recast in a more fruitful, more active light. Paola Cavalieri's collection of essays on animal liberation is an exemplar of precisely this philosophical vantage point on politics. For her, as well as the contributors that she has assembled, philosophy plays a central role in illuminating the moral dynamics that circumscribe our relationship to nonhuman animals and the ways that this particular moral-philosophical problem demands a practical kind of solution.

The relation of nonhuman animals and humans is and has generally been, as Cavalieri makes clear, one of domination and brutality. The philosophical search for moral grounds for the liberation of nonhumans from forms of exploitation, domination, and brutality, however, is unique in this instance since the very essence of the human-nonhuman relation demands that humans themselves reform their practices and commitments. The unique nature of the moral-philosophical problem that confronts us with the conditions of nonhumans therefore makes it incumbent on us as ethical agents to provide the practical context necessary to end the human exploitation and domination of nonhumans. Where Cavalieri and her contributors break new ground is in grasping that the nexus of philosophy and politics needs to be rethought and reconstructed in order to transform social practices.

Cavalieri places social movements at the center of an approach that links this moral-philosophical discourse with a distinctly political one. This is because social movements represent a kind of practical, active instantiation of the changing norms and values of a community as moral contradictions and pathologies come more to the fore of collective consciousness. Philosophy's role here is not only to manifest the moral problems that the abuse and domination of nonhumans poses, but also to show that philosophy itself can no longer be confined to an academic, professionalized role. Instead, it morphs into something political once we recognize that there exists a kind of *practical commitment* that follows once we see that resolving the moral problem of our treatment of nonhumans is premised on the transformation of our distinctively human *practices*. The emancipation of nonhuman animals from human abuse and dominance therefore must lead us to a new synthesis of philosophy and politics, a new kind of practical rationality that demands the transformation of everyday life, social and personal practices, as well as the institutional dimensions of society.

The essays gathered here therefore not only present a series of new, fresh perspectives on the current boundaries of discourse on nonhuman liberation from human dominance, they also operate within a distinctive space of reasons. As a foray into the domain of philosophy and politics, it is a stimulating and morally compelling series of essays that demand that we push against the limitations of our own moral categories and our reified understanding of politics, of philosophy, and of the moral importance of the nonhuman world.

New York City

Michael J. Thompson
Winter 2015

ACKNOWLEDGMENTS

I wish to thank Franco Salanga for his precious cooperation and Harlan B. Miller for his criticisms and suggestions. I am also grateful to Elaine Fan and Chris Robinson from Palgrave Macmillan for their constant help. Finally, special thanks go to Michael J. Thompson, the editor of the series, for his interest in this project and to Tzachi Zamir for his support.

CONTENTS

Notes on Contributors

Elisa Aaltola is a philosopher who has been working on animal issues throughout her research career. She has written over 35 peer-reviewed papers on animal philosophy and has published a number of books on the topic (e.g. *Animal Suffering: Philosophy and Culture*, 2012, and *Animal Ethics and Philosophy: Questioning the Orthodoxy*, co-edited with John Hadley, 2014). She works as a Senior Research Fellow in Philosophy at the University of Eastern Finland.

Matthew Calarco is Associate Professor of Philosophy at California State University, Fullerton, where he teaches courses in Continental philosophy, ethics, and social and political philosophy. His research lies at the intersection of animal studies, environmental studies, and radical social justice movements. He is working on a project entitled *Altermobilities: Profaning the Streets*, and his most recent book is *Thinking Through Animals: Identity, Difference, Indistinction* (2015).

Paola Cavalieri is the founder and former editor of the international philosophy journal *Etica & Animali* and has published extensively on the topic of the moral status of animals. She is the author of *The Animal Question. Why Non-Human Animals Deserve Human Rights* (2001) and of *The Death of the Animal. A Dialogue* (2009). She co-edited, with Peter Singer, the award-winning book *The Great Ape Project: Equality Beyond Humanity* (1993).

Sue Donaldson is a research associate in the Department of Philosophy, Queen's University, Canada. She is the coauthor, with Will Kymlicka, of *Zoopolis: A Political Theory of Animal Rights* (2011), which was awarded the 2013 Biennial Book Prize from the Canadian Philosophical Association, and has been translated into German, Japanese, Turkish, French, Polish and Spanish. She is also a founding member of Queen's Animal Defence, a group working for social justice in a multispecies world.

Will Kymlicka is the Canada Research Chair in Political Philosophy in the Philosophy Department at Queen's University in Kingston, Canada, where he has taught since 1998. His research interests focus on issues of democracy and diversity, and in particular on models of citizenship and social justice within multicultural societies and interspecies relations. His books include *Multicultural Citizenship* and *Contemporary Political Philosophy*, and he is the co-author, with Sue Donaldson, of *Zoopolis: A Political Theory of Animal Rights* (2011).

Brian Luke, formerly Assistant Professor of Philosophy at the University of Dayton, is currently Director of Worship Arts at the First United Methodist Church in Marysville, Ohio. Together with political theory, his main area of philosophical research is the interconnection between gender theory and animal rights, a subject on which he has published essays in collections and journals such as *Between the Species* and *Environmental Ethics*. He is the author of *Brutal: Manhood and the Exploitation of Animals* (2007).

Gregory Smulewicz-Zucker is the managing editor of *Logos: A Journal of Modern Society and Culture* (www.logosjournal.com). He is the editor of *Strangers to Nature: Animal Lives and Human Ethics* (2012) and the co-editor, with Michael J. Thompson, of *Radical Intellectuals and the Subversion of Progressive Politics* (2015).

Dinesh Joseph Wadiwel is Lecturer in Human Rights and Socio-legal Studies and Director of the Master of Human Rights at the University of Sydney. His research interests include sovereignty and the nature of rights, violence, race and critical animal studies. Wadiwel is the author of the monograph *The War Against Animals* (2015) and has over 15 years of experience working in civil society organizations designed to promote anti-poverty and disability rights.

Introduction

Paola Cavalieri

I

When does a political movement really begin to exist? When its core ideas start to appear in a scattered way in the works of different thinkers? When for the first time organized groups begin acting on such ideas? Or when theory and practice merge to create a unified whole able to decode reality with the aim of changing it?

Theory, chiefly in the form of philosophical approaches, has always played a role not only in those recent phenomena we refer to as "movements," but also in most great historical conflicts. This has mainly happened at the level of the regulative ideas giving substance to the ongoing struggles and of the key interpretive concepts adopted. Montesquieu's vision guided the American revolutionaries just as Rousseau's outlook influenced the French ones, and Hegel's theory of alienation was central to the Marxist critique of capitalism. More recently, the philosophical refinement of the egalitarian arguments lent theoretical support to the civil rights movements' demands, while deep ecology discourse, with its attribution of intrinsic value to

P. Cavalieri (✉)
Etica & Animali, Corso Magenta 62, Milano 20123, Italy
e-mail: apcavalieri@interfree.it

P. Cavalieri (ed.), *Philosophy and the Politics of Animal Liberation*,
DOI 10.1057/978-1-137-52120-0_1

1

ecosystems or their components, gave lifeblood to environmental activism. And, obviously, many political achievements were only made possible by theoretico-critical work on accepted concepts and classifications—think of the undermining of the rhetoric of the unnaturalness of homosexuality or of the deconstruction of binary oppositions such as masculinity/femininity or civilization/primitiveness.

All this is rather obvious, and is confirmed by the fact that attempts to destroy movements usually start from the endeavor to destroy their intellectual basis. There is, however, another level on which philosophy can affect the life and growth of movements. It is the level of concrete political choices—of strategic orientation and policy formulations. This aspect is not only less evident, but somehow more controversial, in view of the fact that some sections of political movements are committed to a kind of anti-intellectualism, according to which one should not devote time and energy to argue and theorize when all what is needed against the sheer power of brutality is action.

No one denies the relevance of militancy and of that immersion into practice that can promote both personal and collective growth as well as teach important tactical lessons—many theorists have emphasized the role of spontaneous, self-determining activism. Yet, on the one hand, in the face of an irrational order, the defamation of abstract thought and the total surrender to "the concrete" amounts to nothing less than the surrender of the critical.[1] And, on the other, history tells a more complex story. For better or worse, great movements have not only been informed by theoretical keys to political planning, but have also been marked by heated debates on the philosophical underpinnings of their specific choices. In other words, they tended to place their experiences into a definite framework, which might help to assess, organize, and empower them.

In the leftist tradition, and especially within the cultivated late nineteenth and early twentieth century's European social democracy, veritable battles were fought around major theoretical stances. Thus, the acceptance or the rejection of the notion of dialectics, with its attendant idea that the negation of the inner negativity of reality must necessarily lead to an overturning of the status quo, determined radically contrasting attitudes regarding the choice between revolution and reformism—the example par excellence being perhaps the attack of Rosa Luxemburg on Eduard Bernstein's revision of the Hegelian grid.[2] Or consider the persistent controversies in the theoretical approach to the state and its role. Seeing the state as an institution whose supreme law is the augmentation of its power led some radical leaders to aim at abolishing it altogether, while considering the state as embodying the unity of individuals in an ethical whole impelled others to aim at reforming it.[3] With some obvious modifications, such contrasts persist within contemporary political discussion, as it was just an emphasis on what one might call the "left hand" of the state[4]—or its progressive side—which induced some feminists to demand forms of regulation such as censorship on pornography, while other theorists, stressing the dangers attending the right hand of the state—its repressive side—argue for the rejection of any legal interventions.[5]

It is this more concrete level at which theory, and in particular philosophy, can have an impact on praxis that forms our present focus.

II

It is a general tendency of newly emerging social movements to recur to the ideologies of the forces that preceded them. Most of the nineteenth century's movements, in fact, spoke the language of the French Revolution. But the animal liberation movement's cause has remained in the dark for such a long time, and its proposals so much outrun all actual experience of recent ages, that it is particularly problematic to borrow from previous strands of thought. The only situation with which an approximate parallel can be drawn is perhaps the one of slavery, thus depicted by Hannah Arendt[6]: "[T]he institution of slavery carries an obscurity even blacker than the obscurity of poverty; the slave … was 'wholly overlooked' … Slavery was no more part of the social question for Europeans than it was for Americans, so that the … question, whether genuinely absent or only hidden in darkness, was non-existent for all practical purposes." Not an encouraging situation.

Yet, there are still possibilities. It is such possibilities that, in an exercise of anticipatory political imagination, this work aims at exploring. For generic interpretive frames can be adapted, forms of organization can be extrapolated, and new syntheses can be attempted. But, first, a brief look at the intellectual history of the animal liberation movement may be in order.

An interesting thing to notice is that, contrary to what one may expect, that element of compassion which seems to create a cleavage between "important" and "trivial" political movements—with the former being based on rational claims and the latter on irrational feelings—is not peculiar to the initial forms of concern for the plight of nonhumans, but was clearly present even at the dawning of the social question. Indeed, "the passion of compassion has haunted and driven the best men [*sic*] of all revolutions."[7] In both situations, only a small step was needed to replace compassion with justice. It is well known that, as regards the animal liberation movement, such a step was taken in the 1970s, during the final phase of the analytic emancipation of ethics from religion and metaphysics. Momentous as it was, both for its lucidity and its radicalism, however, and despite the attempted association with other radical causes, especially those of the contemporary civil rights movements, this turn remained mainly an academic undertaking. True, several books and journals succeeded in partially bridging the gap between the new "animal ethics" and the existing, or budding, forms of animal activism, but there was nothing like, for example, the penetration of the social question into the working class of the past two centuries. For one thing, the discourse on animals tended to be relegated to the realm of the so-called practical ethics, rather than investing all moral and political philosophy; for another, analytic authors, though being extremely radical in the theoretical area, were not always politically equipped to develop a critique of established social institutions. Thus, after a brilliant beginning, which stirred some debate, the question ended up becoming the topic of very speculative and abstract disputes, and the animal ethics wave started to recede. Then, after a period of quiescence, the subject matter of animals reappeared in different realms, first in the form of animal studies, and then in the form of critical animal studies. A common feature of these fields is an expansion of the discourse both at the level of the involved disciplines, which range from literary studies to eco-criticism to feminism and gender theory, and at the level of the focal issue, embracing in general animal-human relationships rather than just animal liberation; to this, critical animal studies add a stronger social commitment and a definite leftist slant favoring a renewed attempt to merge the cause of animals with other social causes. Finally, in the last period, one can see the rise of a new and focused field of reflection in which political philosophy and thinking directly enter the scene of discussion on nonhumans.

This field is still heterogeneous. On the one hand, it keeps including conventional reformist discourses that basically accept the status quo. On the other, however, it comprises an area of research which, embracing in a radical vein the antagonistic, rather than reconciliatory, dimension of the political,[8] articulates new conceptual schemes and pragmatic avenues, to the effect that, in a moment when many traditional intra-human movements waver, the animal liberation movement starts to constitute itself as a structured social reality. It is within the context of this latter area that the present work develops a collective investigation of the form(s) that a politics of animal liberation might assume, with an eye to introducing into the discourse the aptitude peculiar to ideologies to pronounce on which conceptual and pragmatic solutions are available for the understanding and shaping of the political world.[9]

The endeavor to confer on a set of different positions the contours of a unified project poses some difficulties. In the first place, while not requiring a strict dialogical form, the effort demands at least some homogeneity. Independent and unconnected views, however illuminating, can hardly give rise to a significant intellectual challenge, or become a catalyst for the articulation of other substantive views. In this light, the contributions of the book, which come from the main areas of the current debate, have been jointly stimulated through the circulation of the opening chapter. The authors have vigorously reacted to the picture offered by this chapter by appealing to their own traditions and methodologies, thus providing a mosaic of pieces which, together, may engender a cumulative work of symbolic construction.

III

Even though a necessary characteristic of any pragmatically oriented philosophical discourse is the quest for a language which avoids excessive technicality, so as to be able to concretely formulate not only abstract ideas still untested by reality but also action-guiding prescriptions,[10] at present this sort of discourse cannot but come from universities. These are no longer the years in which radical intellectuals spent their life in study and reflection "in famous libraries of London and Paris, or in the coffee houses of Vienna and Zurich."[11] Critical thinking is now cultivated in the academy. But isn't this a problematic site? Aren't universities prima facie, as parts of the global education system, disciplinary institutions meant to be integrated in, and supportive of, the status quo,[12] where, moreover, cultural competition may distort the actors' interests, favoring the pursuit of uncontroversial topics and the production of commonplace results? Undeniably so. But first, we know that there may be exceptional conditions in which external outbursts of social critique can, and do, find in the academy a natural theoretical expression. And second, universities may become the stake of ideological power struggles, thus opening the door to the production of theoretical instruments which can be appropriated and refined by other social agents.

Actually, the fundamental problem posed by a scholarly matrix for a political discourse is a more theoretical one, which can be summarized by reference to the critical notion of "scholastic point of view." This notion, which, in its first formulation, was meant to denounce the distortions that idiosyncratic philosophical usages can induce in the treatment of philosophical problems, was subsequently given a more political twist, which, while extending the critique to wider contexts, raised the question of the social conditions of possibility, and of the political-intellectual consequences, of such a peculiar viewpoint.[13] The particular aspect that interests us here is the one concerning the freedom from necessity and urgency that is the normal condition for the academic exercise as a gratuitous game, liable to raise speculative problems "for the pleasure of resolving them," and not because they are posed, often quite urgently, by real-life situations. For how might one accept such a neutralizing disposition in matters of politics, and especially of a politics whose center of gravity is the most neutralized among oppressed groups?

This is a serious problem for all movements. But, in a sense, it is just the particular nature of the oppressed group in question that makes it impossible for the intellectual advocates of animal liberation to fully withdraw from the world of action. For nonhuman exploitation is so pervasive and widespread that all concerned are immediately confronted with the necessity, out of mere consistency, of rejecting the social paradigm and the practices it licenses by changing their basic life habits. This change involves a form of *practical* commitment which cannot be found in most other political-intellectual fields, and which prevents the possibility of a totally detached, objectifying approach. Thus, it can be claimed that animal liberation philosophy is presumptively less affected by what is perhaps the most negative implication of the scholastic view.

IV

As noted above, the contributions to this volume come from quite different backgrounds. Thus, it can be helpful to offer a sort of aerial view which might frame them starting from their authors' cultural matrix and central conceptual tools.

After the sort of introduction represented by my chapter which, within a critico-dialectical approach, essentially focuses on the search of a perspective in the literal sense of a theoretical vanishing point whose effect of convergence might give unity to dispersed political activities, Matthew Calarco opens the discussion with a direct challenge to shared assumptions in animal ethics. A versatile scholar, whose reflections range from animal and environmental ethics to continental philosophy, with a special interest in French Theory's deconstructive bent, Calarco develops in particular an attack on that notion of speciesism which, since the start of the analytic debate, has been a powerful instrument for the critique of the human-ist paradigm because of the immediate parallel with racism and sexism it establishes.

The basic theoretical tool through which such an attack is waged is the concept of anthropocentrism—a concept that has known great popularity in environmental ethics, but that, in Calarco's perspective, takes on addi-tional, more complex meaning. The feature of the notion that is stressed by both the old and the new construal is the idea that human beings are the most significant entities in the world. Thus, Calarco clearly states that anthropocentrism is premised chiefly on the belief that the human is exceptional among animals and all other beings. To this, however, in consonance with some radical strands of feminist thought, he adds the view that anthropocentrism embodies a network of "ideas, structures, and practices aimed at establishing and reproducing the privileged status of those who are deemed to be fully and quintessentially human"—that is, not of the human species as such, but of a small subset of beings belong-ing in the species.

A corollary of this construal is that the dominion over, and the exploi-tation of, nonhuman beings, though fundamental, is embedded in a web of oppression also embracing various kinds of discriminated human indi-viduals or groups. And the political implication of such a discourse is that the struggle for animal justice cannot be fought in an independent way, but must instead draw force and insights from the cooperation with other radical social movements. Thus, if the animal movement's theory must be inter-disciplinary and inter-sectional, simultaneously working along mul-tiple lines of analysis, the movement's strategy of resistance should focus on interlocking systems of power, and should therefore include among its tactics an involvement in those alternative minoritarian cultures which strive to overcome productivism and commodification by contesting the practices that negatively impact animals, humans, and the environment.

Such an attention to other movements for social justice is shared by Sue Donaldson and Will Kymlicka, who so deeply contributed to the recent political turn in animal philosophy. This should come as no surprise, as the authors of *Zoopolis*, besides introducing the reflection on nonhumans into the political field, directly inserted contemporary political theory into the discourse on animals, thus exploring new possible intersections as well as new normative proposals for a society that might include nonhuman subjects as citizens.

In their contribution, Donaldson and Kymlicka extend such an approach in the direction of the development of a praxis for social change concretely shaped by Rawls-inspired principles rooted in the leftist, egalitarian wing of contemporary liberalism, and strongly critical of the burgeoning rightist libertarian dogmas. In doing this, they produce an elaborate discourse, which includes a wide-ranging *pars destruens* and a detailed *pars construens* embracing both general strategic suggestions and concrete examples. Foremost among the targets of their critique of the present state of the animal movement are the excessive focus on individual conversions and personal ethical transformations, the tendency of some advocacy organizations to act like business corporations, and the self-closure of a political attitude that appeals to "moral baselines" that erect barriers toward possible allies. According to the political dynamics Donaldson and Kymlicka envisage, in contrast, successful reform requires an organized movement which, starting from a reshaping of local communities and employing a bottom-up procedure, engages in the process of institutional change in cooperation with other progressive causes and as an integral part of the broader social justice movement. In consistency with this framework, the definite suggestions advanced to offer a glimpse of a possible renewed activism range from the integration of nonhumans into the urban areas designed for universal access to the creation of interspecies residence-work-care models to the fostering of place-based, progressive animal-friendly economies.

However complex, such a discourse is manifestly dominated and unified by a specific theoretical model: that of community building. It is just such a political enterprise which, by creating and diffusing new lived environments—environments which link animal defense to ideas of democratic renewal and protection of the vulnerable—may lead to fundamental changes in power relations and, in the long term, to the creation of a global interspecies political community.

The relevance for the thriving of such communities of felt experiences and lived social environments leads to the problem which is more generally taken up by Elisa Aaltola in her analysis of akrasia, that is, of that state of weakness of the will in which one does what one rationally recognizes to be wrong. How, Aaltola asks, can the animal movement deal with this stumbling block which lies in the way of nonhuman emancipation, especially under the guise of what she defines as the omnivore's akrasia, or the attitude that allows the perpetuation of a mass participation in the exploitation of animals in its main form of raising and killing them for food?

Aaltola's Aristotelian bent shows in a long-standing interest in the non-strictly rational aspects of morality, the ethics of virtue, and the centrality of empathy, with that attendant focus on moral psychology which comes to the forefront in her contribution. In this context, however, Aaltola surpasses previous reflections of like-minded authors such as Mary Midgley since, with a radically political shift, she expands her analysis to include components external to the individual mind. Drawing from philosophical as well as scientific explorations of weakness of the will, she argues that internalist explanations focusing on individual egoistic desires or muddled emotions miss the point, and that akrasia can be fully understood only in reference to the surrounding societal milieu. In the case of the omnivore's akrasia, in particular, the attitude's sources must be sought in an economic and political environment whose institutions and customs continually instigate animal consuming habits.

An externalist approach clearly changes the nature of what is traditionally seen as the main remedy to akrasia, that is, self-cultivation. Within a politically aware stance on akrasia, Aaltola argues, this primary concept must be enlarged into what one might define as societal cultivation—a form of cultivation that is favored and enacted by a morally sound social environment. Given that, however—with the exception of a utopian Aristotelian world—such a form of encompassing cultivation can only be achieved through a prior, difficult reshaping along those lines of contemporary society, what we face is a conclusion urging the animal movement to switch from a politics concerned with personal choices to a politics of confronting, and challenging, social structures.

Up to this point, the discussion has been manifestly political. The chapter which follows, by Brian Luke, adopts instead a cultural critical reading, focusing on intellectual heritage and mythical-ideological narratives. As Luke makes clear, though, this narrative approach remains deeply political, "in the anarchistic sense of recognizing that radical social change is neither initiated nor consolidated by the state." Indeed, according to Luke, legislative reform is itself, among other things, a form of narrative, as are other kinds of governmental action.

Coming from the analytic tradition, contributor to such pioneering publications as *Between the Species*, with a penchant for care ethics gradually developed into the adoption of the radical feminist theory, Luke offers an analysis of the logic of human domination (at both the intra- and inter-specific levels) through a survey of four texts that contributed to shape Western consciousness according to the paradigm of a layered system of exploitation and that, while centering on humans, also include references to animals: the myth of Iphigenia and children's ritual sacrifice, the Shakespearian parable of the "shrew" and the subjugation of women, uncle Tom's story and American slavery, and Sinclair's account of labor abuse in early twentieth-century slaughterhouses.

In each case, the result of Luke's analysis of the connections between forms of intra-human oppression and the oppression of nonhumans is that animals are beings whose exploitation is so natural that either it goes unnoticed or it is simply present as the negative term of comparison. All the more so: a review of some modern attempts to soften, or to reinterpret, the less palatable aspects of the narratives in question shows that they concentrate on human exploitation, thus demonstrating that the abuse of the members of other species still remains "the default model for describing the oppression of a human group." In the light of this, how can one move forward? Luke's response revolves around the device of *retelling*, or of creatively intervening through variations in old, influential tales: to counter the institutions of nonhuman exploitation, and to denaturalize and rehistoricize the phenomenon of the human domination of animals, in addition to more straightforwardly political efforts, we must also be engaged narratively. In other words, Luke identifies a chief element of our opposition to animal exploitation in the act of taking on our culture's stories, bound to uphold an anthropocentric worldview, and of turning them into something radically different, productive of a new zeitgeist.

Unlike Luke's approach, Dinesh Wadiwel's stance is inserted in a clear-cut categorial grid. For Wadiwel works within the frame of that biopolitical discourse that, initiated by Michel Foucault, has articulated a view according to which the essence and/or the developments of Western politics revolve not around the traditional liberal idea of agreement, but around the gradual sovereign subjugation of humans not only as political beings but also as mere living things.

Wadiwel contends that war is the concept which best captures the nature of our constant, structured violence toward animals. It is just the continuing victory in this unbalanced war against resisting but weaker beings, not any preexisting superiority, that grants humans that sovereignty over nonhumans, which allegedly warrants a right to use and consume them as an exclusive form of property. In such a context, Wadiwel observes, it cannot be surprising that that form of governmentality—in Foucauldian terms, the management of populations and individuals through techniques and dispositifs that completely capture life—whose extreme intra-human manifestations are concentration camps or detention centers, has its origin and paradigmatic manifestation in the institutional sites of violence where animals are literally kept at the threshold of life and death.

Is there any chance to challenge this integrated oppressive system? Wadiwel makes reference to two political tools. One is counter-conduct, that is, a sort of desertion or insubordination against that which exists, an instance of which, in the case of nonhumans, is the practice of veganism not as a form of personal asceticism but as a means to disrupt systemic violence and a mechanism for building alternative communities. The second, and more important tool, is the hypothesis of a truce—one day without killing animals, one day when the slaughterhouses would shut down, and when we might imagine a space for a different ethical relationship. Wadiwel admits that this strategy, even as a thought experiment, may seem ambitious. However, he underscores on the one hand the fundamental symbolic relevance of focusing not on consumers, but on the production process of killing animals, thus operating explicitly on an institutional level, and on the other the fact that, in this attempt to temporarily disarm sovereignty, animal advocates might adopt an alliance approach and look for support from other transformational movements, and in particular from those workers who are the exploited section of the exploitative system.

Though founded on highly theoretical premises, it is just the problem of the realizability of the movement's political goals that lies at the core of the chapter by Gregory Smulewicz-Zucker. For, with this last contribution, we go straight to the question of movement building. Actually, Smulewicz-Zucker is not only involved in the animal question, but has also a background in intra-human politics, and his stance is fundamentally the same in both realms: he is interested in a progressive political discourse relevant to the machinations of real politics, in a rational radical stance which might produce a critical confrontation with asymmetrical power relations.

Like Donaldson and Kymlicka, Smulewicz-Zucker espouses a form of left liberalism aiming at the democratization of social life; unlike them, however, as his critique of liberal approaches resting on theories of deliberative democracy shows, he incorporates a strong Hegelian influence in the form of the idea that rights only become actual when they are institutionalized by the state, and that political movements are a means of pressuring the state to establish and extend rights. Manifestly, such a defense of the role of the state as a possible ally in the struggle for intra- and inter-specific justice—a defense that somehow reactivates the well-known controversies—with the associated idea that it is the law which can drive the rhythm of social progress, prompts him to reject not only those versions of liberalism which want to minimize or strongly limit forms of state intervention, but also the theoretical perspectives which he sees as marked by direct anti-statism, such as neo-anarchism, or even by a total lack of political direction, such as postmodernism and posthumanism.

What does Smulewicz-Zucker think a more effective animal rights politics would look like? After an extensive, though sympathetic, critique of the present condition of the animal movement, its organization, and its tactics, he proposes a shift which, while embracing various specific tactical suggestions, fundamentally revolves around the key notion of *pressure*: all the factors at play, whatever the differences between them, must cooperate to pursue one single aim—the aim of compelling the state to grant legal rights to nonhuman beings. Thus, through the focus on a strategic line which is distinctly centripetal, Smulewicz-Zucker defends a relevant alternative to those centrifugal perspectives that privilege public involvement and outreach programs.

Judging from the above it seems fair to claim that all the chapters in question clearly testify to the role that philosophy can play in fostering innovative visions and practices.

NOTES

1. See Herbert Marcuse, *Reason and Revolution: Hegel and the Rise of Social Theory* (Amherst, NY: Prometheus Books, 1999), 370.
2. See Kurt Lenk, *Teorie della rivoluzione* (Bari: Laterza, 1976).
3. For the former position, see, for example, Mikhail Bakunin, *The Selected Works of Mikhail Aleksandrovich Bakunin* (Los Angeles, CA: Library of Alexandria, 2009), 170; for the latter, see Ferdinand Lassalle, "The Worker's Program," in *Socialist Thought*, ed. Albert Fried and Ronald Sanders (New York: Anchor, 1964), Chap. VIII.
4. For the distinction see Pierre Bourdieu, "The Left Hand and the Right Hand of the State," *Variant* 32 (2008), http://www.variant.org.uk/32texts/bourdieu32.html.
5. See Catharine MacKinnon, *Only Words* (Cambridge, MA: Harvard University Press, 1993), 15ff. and Judith Butler, *Excitable Speech: A Politics of the Performative* (New York: Routledge, 1997), 82ff.
6. Hannah Arendt, *On Revolution* (London: Penguin Books, 1990) 71-72.
7. Ibid., 71.

8. For the distinction between the dimensions of the political, see Chantal Mouffe, *On the Political* (London and New York: Routledge, 2005), "Introduction."

9. On these features of ideologies see Michael Freeden, *Ideologies and Political Theory* (Oxford: Oxford University Press, 1998), "Epilogue."

10. There are possible models. The most interesting comes from a series of conversations following two lectures held by Herbert Marcuse in 1967 at the Freie Universität of Berlin, which directly dealt with topics like the individuation of political turning points and with pragmatic concepts such as those of latent potentialities. See Herbert Marcuse, *Five Lectures: Psychoanalysis, Politics, and Utopia* (Boston: Beacon Press, 1970), 62–82.

11. Arendt, *On Revolution*, 259.

12. For universities as "disciplinary blocks," see, for example, Michel Foucault, *Discipline and Punish* (London: Penguin, 1991), 138.

13. For the first formulation of the notion see J.L. Austin, *Sense and Sensibilia* (Oxford: Clarendon Press, 1962), 3; for its expansion see Pierre Bourdieu, "The Scholastic Point of View," *Cultural Anthropology* 5/4 (1990): 380–391, and id., *Pascalian Meditations* (Stanford, CA: Stanford University Press, 2000), Chaps. 1 and 2.

Animal Liberation: A Political Perspective

Paola Cavalieri

All animals are equal.
   Peter Singer[1]
   ... To draw up a plan of action, be it violent or non-violent (because against social injustice all means are legitimate), by which we would contribute to the destruction of social fictions... and we would already create, if possible, a little of that future freedom.
   Fernando Pessoa[2]

In a world ideologically dominated by humanism—the socially inculcated doctrine of human superiority warranting the discrimination against, and the exploitation of, all the other animals—where capitalism as a system driven by an impersonal logic of accumulation, appropriation, and commodification has achieved the complete reification of nonhuman beings,[3] what are the political prospects for a movement for the liberation of animals?

While animals have been tortured and killed en masse during all human history, nothing that happened in the past can equal their present treat-

P. Cavalieri (✉)
Etica & Animali, Corso Magenta 62, Milano 20123, Italy
e-mail: apcavalieri@interfree.it

© The Author(s) 2016
P. Cavalieri (ed.), *Philosophy and the Politics of Animal Liberation*,
DOI 10.1057/978-1-137-52120-0_2

ment. On the one hand, despite any attempt to dissimulate the situation on the part of official discourse, the legacies of the past have not disappeared, but have simply been excluded from view by relocation and concealment. On the other, the spread of industrialized methods of generating, raising, sustaining, and killing for food or research, coupled with the ever growing scientific and technical dominion of organisms, generated an exploitation system without parallels. For such a system not only involves unprecedented numbers of victims, but also implies such invasive practices as genetic manipulation, artificial insemination and embryo transfer, cloning and pharmaceutical enhancement. Indeed, if modern totalitarianism is the realm where "everything is possible,"[4] nonhumans constantly live under totalitarian rule. At the material level, exploitation keeps growing, with the industrial explosion in formerly undeveloped countries. At the symbolic level, even the rare critical voices that, within philosophico-political discourse, are raised against the direct subjection of biological organisms to political sovereignty and technological apparatuses not only ignore the condition of other than human beings, but, through the indictment of dehumanization, even implicitly ratify it.[5]

Neither humanism nor capitalism can be the immediate target of political action. As has been noted,[6] in politics, one does not oppose anonymous systems, but concrete social agents and their agendas, and concrete institutions and their policies. This means that strategy is at the core of any political movement. Unlike tactics, focused on the short-term means used to reach a specific goal, strategy is addressed to the basic logic of the system to be attacked, and must offer a distinctive, internally coherent, and analytically inspirational theoretical perspective.[7] Starting at least from the end of the 1980s, the animal liberation movement has produced reflections on strategic questions. Since, however, such reflections tend to cluster around the extremes of either adopting largely practical procedures that merge the descriptive and the normative, or offering outlooks that are too general to provide articulated instruction,[8] it can be worthwhile to consider the question afresh, taking as a starting point a survey, and critique, of the two currently prevailing approaches to political activism.

BETWEEN LIBERALISM AND THE LEFT

At present, important sections of the animal liberation movement think and act—most often with passwords borrowed from discussions in analytic moral philosophy—within the framework of a liberal perspective. And this not because contemporary political liberal thinkers have paid particular heed to the predicament of animals, but because a liberal stance instantiates conformity to the given. Liberalism is the water in which we all currently swim, and to act according to its rules is the most natural thing to do.

As far as the problem of socio-political change is concerned, contemporary liberalism takes democratic institutions seriously: decent societies can be improved through gradual changes, which do not normally involve insurgence or violent actions, but rely instead on the gradual illumination of minds and on the recourse to democratic procedures and incremental legal reforms.

Sometimes, exponents of the animal liberation movement put the accent on the former course. For instance, Gary Francione has advanced the proposal that the international movement should arrange "a sustained and unified campaign promoting a purely vegetarian diet," making people aware of why they should not use animal products at all, and educating the public about the need for the abolition of animal exploitation.[9] Slightly different, but related, is the approach that sees the activist as anyone who goes into the marketplace with a dollar in hand, and says "I'm going to buy this rather than that because it has something to do with the way that animals are treated."[10]

What those operating from within such a framework attempt is to directly convert each and every individual to respect for animals. By selectively focusing on this undertaking, however, they find themselves in a difficult situation. Not only must they face the fact of the instinctive prima facie opposition of a large section of the population. Not only do they engage in an endless enterprise which, like writing in the sand, is always subject to reversals and backlashes (something that holds as well for straightforward attempts to influence people through the results of undercover investigations, or to change individual corporations' production practices). But, more fundamentally, they appear to misinterpret our world. Addressing people (or even social entities) in an unmediated way means overlooking the fact that we do not inhabit simple societies whose habits are shaped by customary rules emanating from group members, but live instead in differentiated societies, whose basic policies have been incorporated in laws and institutions. It is such objectified fruits of group normativity, and the exploitive practices they endorse, that must substantially be targeted if one wants to alter the situation.

Aware of this, some liberally minded intellectuals and activists focus instead on the latter course, turning to legal reform. However, first, given their basic acceptance of the social framework, they do not feel the need to develop political analyses and carefully worked-out plans, often being content with either scarcely controversial or momentarily popular issues.[11] Second, fearing the hegemony that excludes dissidents from the possibility to speak, and also, in some cases, the loss of an officially recognized status which facilitates both economic and social survival, they are inclined to shift from radical claims to requests for moderate measures that do not challenge the status quo. While such measures can indeed reduce animal suffering—and, as has been stressed, the current plight of animals is so severe that "any reduction in the hell they endure is laudable"[12]—they are already pursued by traditional animal protection organizations. It makes thus little sense that they are included among the goals of the animal liberation movement.

Perhaps because of the problems with both the "conscious consumer" view and the random legal approach, a growing part of the animal liberation movement is turning its attention to that Leftist galaxy that adopts a more questioning attitude toward social reality and its structures, focusing on political exclusion and economic imbalances, and on the play of race, gender, or ethnicity in social order. Matthew Calarco well epitomizes this stance: "It is becoming increasingly clear that, if the animal defense movement is to be more effective politically, it is going to have to find ways to link itself with other progressive movements for social change."[13] In this case, what one is confronted with is not so much the production of a detailedly different set of forms of activism, but rather, given the nonconformity of the choice, the attempt to make room for the struggle for nonhumans within the existing web of anti-system approaches and struggles.

Prima facie, this does not seem an easy avenue. For those who resist discrimination and ideologically oppose global capitalism, tend to pay no attention to the animal question. Indeed, what can be still called the Left is committed to humanism. With very rare exceptions, leftist intellectuals simply erase nonhuman beings. And though it is of course difficult to point to the lack of something, there are, however, particularly relevant loci where the erasure becomes blatant.

Thus, one finds no mention of animals within discussions of a universal which becomes articulated through challenges from those "who are not covered by it"—the excluded constituting the contingent limit of universalization are merely *human*.[14] Even more clearly, in the context of a reflection on the forms of domination to be challenged, following a definition of democratic revolution as the "progressive extension of the principle of equality," the history of such extension is unquestioningly characterized as the history of *human* universality.[15] On the other hand, in the rarest cases in which the issue of nonhumans casually surfaces, one meets with revulsion for the possibility of erasing the boundary between humans and animals, or of leveling the animal/human divide, as this might allegedly contribute to giving "the green light to all sorts of human rights abuses."[16] All in all, the Left whose support is envisaged as fundamental appears unable to acknowledge the seriousness of the transformation of the political agenda that the issues posed by the new animal politics imply.[17]

Just as most mainstream political thinkers, Left theorists succumb to that form of conditioned ethical blindness which is perfectly epitomized by the commonplace idea that *cattle railcars* are inadmissible for humans, but perfectly admissible for *cattle*. Thus, while keeping animal carnage at a distance, they not only reflect the exclusionary attitude that has always informed the defenses of intra-human oppression, but also foster the caste egoism which often led the aristocracy of labor to turn blind eye to the fate of the coolies of the earth.

And yet, since alliance building is important, the Left continues to hold its fascination. For the question remains: in a world where in most nations exploitation and repression are routine, and in a few luckier ones welfare policies are gradually eroded,[18] why not try to develop political connections by standing beside those post-Marxist movements which try to defy the existent order? The umbrella, though small, has a dignified tradition: after all, the success of Marxism displayed the greatest theory effect as ever by successfully realizing its potential in the social world. Why not, then, profit from it, by joining in its struggles, while in the meantime helping it overcome its humanism?

The problem with this approach, favored in the first instance by intellectuals working within socialist frameworks,[19] but also popular within French Theory and post-humanism, especially under the auspices of the continental strands inspired by Michel Foucault, Gilles Deleuze, and Jacques Derrida,[20] is that it too easily overlooks the ideological pitfalls of the social struggles it embraces. For, as could be expected, movements grown out of essentially humanistic concerns simply overlook the theme of animals, at best attacking factory farming for fear of monoculture, and at worst focusing on aboriginal rights to hunt and trap.[21] Despite the fact that such ideological problems are not always appreciated by continentally oriented approaches, due to the looseness of their ethical grid,[22] it is clear that if the animal liberation movement insists in pursuing this path, just as with liberal activism, it can merely produce requests for moderate measures that do not defy the status quo.

LIBERALISM AND THE QUESTION OF CITIZENSHIP

There are further avenues, however. Whatever defects "really existing liberalism" may have, radical liberal thinking can be surprising. And, sometimes, it can reach heights that socialist traditions can hardly conceive of. The attribution to nonhumans of basic human rights is not enough, Sue Donaldson and Will Kymlicka argue, envisaging a future Zoopolis. For what animals need is political recognition—citizenship if they live among us, sovereignty if they live in their own territories, and a form of denizenship or partial citizenship if their communities overlap with human communities.

Nonhuman citizens, whose basic negative rights would be legally enforced and who would be represented through forms of dependent agency in political decision-making, should be facilitated in the realization of their own life projects, and should also learn the responsibilities incumbent on members of society. Sovereign animal communities, which can collectively respond to the challenges they face, and can provide a social environment where their members might flourish, should be protected from human harms and should obtain a secure space safeguarding their autonomy against external threats of annihilation, exploitation, or even assimilation. Finally, animal denizens, often having no other place to live, should have the right to secure residency in urban areas, where their interests would be taken into account in human decisions, and their right to anti-stigma measures would be backed up by full protection of the law.[23]

Is the concept of a Zoopolis a utopia in the classical sense? In one respect, yes, for it radically challenges existing conditions which are a long way even from its necessary prerequisite, that is, an extension of basic equality to animals. But utopias play an important role in history, offering regulative ideas that guide civil progress. The force of utopias is so great that their proponents were often persecuted, and the continued success of socialism in keeping social conflict alive is in part due to its utopian horizon of expectancy—the dream of a totally just society to come.[24]

In another respect, however, the reference to the role of forms of political recognition in granting protection to the powerless is far from eccentric, being rooted in a tradition that acquired special relevance within the reflection on recent intra-human tragedies. In particular, while discussing the role played by stateless people in the emergence of the twentieth century's totalitarianisms, Hannah Arendt argued that universal human rights do not work precisely when they are most needed, and that only political recognition, in the form of membership in a particular state, can offer a shield in the world as we know it.[25] Beings deprived of political recognition from existing political communities are within society without being part of society, as they lack the "right to have rights,"[26] and remain thus liable to be coerced or killed en masse. In the light of this, and of the fact that, for radical liberal thinking, political communities are never a completed totality—there is no ideal democracy, but only processes of democratization[27]—it is not surprising, even in a period when the doctrine of human rights has developed well beyond anything seen in Arendt's times, to assist the continuous resurfacing of the question of citizenship, even in once totally unforeseen realms.

In fact, of the three requests advanced by Donaldson and Kymlicka, citizenship can be seen as the strongest from a symbolic point of view, but the most controversial with regard to actual enfranchisement. And this because, unlike sovereignty and denizenship, which imply looser ties, citizenship sanctions forms of human/nonhuman coexistence which can become risky for the much weaker nonhumans. Here, a policy of gradual extinction of domestication seems undoubtedly safer. But, given the importance it takes on in the barrier that Arendt and, partially under her influence, many political thinkers erect around *human* life, citizenship has certainly a role to play in the political discourse about nonhumans. Perhaps—and we shall come back to this—one might select it as a solution for especially difficult situations.

On the other hand, if nonhumans are owed forms of political recognition, how might one turn the polis into a Zoopolis? Or how can one at least come closer to this end, taking the preliminary step of removing animals from their condition of social nonexistence? Applying to the nonhuman case the full apparatus of contemporary liberal political philosophy is a lesson in universality and consistency. But what about the road to follow to achieve change—what about social agents, power relations, and strategies? Is liberal theory as able to envisage the means as it is to suggest the goals?

As mentioned, liberalism focuses on the recourse to democratic procedures and gradual reforms. This is an especially awkward path when what is at stake is radical change. After all, as Max Weber suggests, it took the French Revolution for the European peasant to cease to be considered merely as a means for supporting some ruling classes.[28] Should one suppose that animals might be liberated from a much deeper exploitation in the absence of an analogous process? True, one could make the move of appealing to self-determining developments: given the unsustainability of the present situation of the planet, and the risks for the exploiters themselves, the inevitable course of history might bring exploitation to an end. But is this a realistic suggestion? Apparently not. Mutatis mutandis, the idea of victory through the inevitable course of history was entertained by Marxism as well, but, as we know, to no avail.

The Left and the Question of Revolution

If history itself does not help, and random reforms seem to be insufficient to achieve profound change, what about revolution? The pendulum goes back to a Leftist approach—but one replacing a politics of micro-insurgency fostering plural narratives within the existing order with a totalizing narrative favoring socialistic or at least noncapitalistic outcomes.[29] The question then becomes: can a section of that Left which is so inadequate with respect to the ends be at least useful as regards the means? To conceive of a revolution—to foresee what Jacques Derrida aptly characterizes as "a radical break in the course of history"[30]—one should perceive established institutions not as perfectible embodiments of acceptable principles, but as the elements of a system which, being the result of power struggles and supporting the more or less hidden agendas of the dominant, cannot be altered, but must be replaced. And replacement implies a radical clash and the seizure of power—something which should be prepared through a serious organizational work.

Such a work, always hard, is even more difficult in a period when global capitalism displays its ability to interpenetrate all existing institutions and to defeat or assimilate opposition. Can the current intra-human revolutionary approaches perhaps offer some cues? This is doubtful. For, undermined by the collapse of the eschatological, or even messianic, structure of the Marxist narrative, and confronted with the fragmentation of the traditional social landscape, such approaches—advanced by intellectuals whose agency appears restricted to acts of theorizing since, unlike most of their predecessors, they are not leaders of political organizations—seem able to produce only dubious political abstractions. As the place once reserved for the main social character—the "authentic" working class—appears now empty, the scene is dominated by heated debates on the possibility to identify new social hegemonies or opportunities. Thus, authors working in a neo-Marxist tradition try to grant a key structural role to the immaterial labor embodied by multitudes composed of diverse political actors moving in a transnational and transclass space[31]: other, more ontologically minded thinkers forebode, in a quasi-theological scenario, the intervention of miraculous Events causing the emergence in history of eternal truths and fostering the restitution of the "inexistent" of the world[32]; and, in others still, with a radical gesture apparently renouncing any actual relationship with the masses, appeal is made to redemptive repetitions of past insurgencies, meant to release their full utopian potential by seizing now their missed opportunities.[33] Furthermore, in their eagerness for actual revolts, such intellectuals may become ideologically compliant, inasmuch as, while the tide of Occupy Movements is rapidly withdrawing, the only insurgencies recently having an impact are located in countries where the political field has not yet fully differentiated from the religious field, and where, as a consequence, the aftereffects can be at most dangerous, at worst harmful, for women, homosexuals, minorities, and the like.[34]

If this is the state of affairs as far as human beings are concerned, one can only wonder what might be the prospects of a revolution on behalf of animals. But let us imagine for a moment that a French revolution for nonhumans is conceivable—that a concrete possibility opens up for a revolt in which either animal defenders themselves are the leading characters, or the insurgents incorporate their goals. There are still problems.

For one thing, revolutions can fail—indeed, most attempts at revolution do. If so, the consequences are usually devastating: ongoing processes of emancipation are brought to a halt or reversed for long, and, while the previous avenues for change are precluded, the very existence of the involved forms of social agency is jeopardized. For another thing, and more importantly, even successful revolutions inspired by great visions of a fairer social order can, after an initial period of reforms, end up producing despotism or highly repressive societies. The "heretics and renegades"[35] among the revolutionaries have advanced a number of analyses of the possible causes of this phenomenon. Among them, internal reasons like the sudden evaporation of a concentrated revolutionary class; external factors like military attacks or the pressure of uneven and combined economic development; and, at a more general level, the autocratic involution of the group which seized the power due to a collapse in the insurrectional participation, the "immaturity" of the masses or of the times, or even human nature itself. Anyway, whatever the reasons, it is undeniable that it often occurs that the liberating purposes sink in the dark as soon as they turn from pure lawmaking revolutionary violence to the law-preserving violence of the state.[36]

Probably, then, one should hope for revolution only when one has nothing to lose—when the situation is so bad that it can only improve. Such is, indeed, the situation of nonhumans in our societies. The only thing the animal movement could lose—apart for a thin humane veneer covering savage exploitation—are the liberating discourses developed in the past 40 years. Are they so precious as to be worth giving up an albeit risky revolution? Probably not, if only because of the importance of a radical breaking off of the status quo. But it's pointless to ask now, in the clear absence of any prospects.

A Way Out?

Perhaps what is needed is an alternative vision—one disenchanted enough to provide effective theoretical guidance into the woof of reality while at the same time not being so discouraging as to prevent political action by a movement whose concrete prospects are especially difficult. And, if one searches for the seeds of such a vision within the tradition of critical thought, one meets with at least two avenues which seem able to offer both epistemic and practical cues: Frankfurt School's Critical theory and Pierre Bourdieu's Reflexive sociology.

The first outlook centers on the idea that capitalism is just one, even if perhaps the most impressive, among the expressions of a more fundamental phenomenon, the dominion of instrumental reason, or of that aspect of Enlightenment rationality which, in its craving for the domination of objects, does not hesitate to turn even subjects into objects. Viewed as "the agency of calculating thought" which apprehends everything in terms of manipulation and administration,[37] and construed as a sort of essence which is both expressed and masked by the different phenomenal forms it takes in various historical contexts,[38] instrumental reason is central to perhaps the most striking *j'accuse* directed at Western civilization, Theodor Adorno and Max Horkheimer's *Dialectic of Enlightenment*.

Actually, what has come to be seen as the magnum opus of Frankfurtian Critical theory has not escaped the attention of some intellectuals in the animal liberation movement,[39] mainly because—*rarae aves* in the Leftist tradition—Adorno and Horkheimer, while still far from envisaging a basic reappraisal of nonhuman moral status, nonetheless grant a relevant place to animal oppression within their socio-philosophical approach to the issue of exploitation, so that the struggle for animal liberation clearly appears as a part of the endless struggle of the dominated against the dominant.[40] In comparing current social structure to a skyscraper whose roof is inhabited by the powerful among humans, whose intermediate floors are populated by ever more wretched individuals, and whose basement is "the animal hell in human society,"[41] Max Horkheimer even seems to assign to nonhumans the role of the most exploited "class"—of the group, that is, which just because of its placement can, though being particular, become the carrier of universality with respect to the social totality.

There is, however, a further aspect which deserves the movement's attention—an aspect having to do with Critical theory's more general stance. The Frankfurt School has been blamed for its negative pathos. But the long shadow of European fascism stretched over the *Dialectic*, and, after the dissolution of revolutionary hopes due to the tragic involution of the Soviet experience and in the face of late-capitalism's celebration of the final "enthronement of the means as the end,"[42] it was difficult to detect any immediate possibility of profound social change. Moreover, after and contra Marx, in whose wake it operated, the School had gradually come to realize the fundamental relevance of the superstructural factor of ideology, no longer seen as a redoubling of reality in the realm of the illusory, but perceived instead as somehow coinciding with reality.[43]

It was from the ensuing discomforting sense of the need for a new approach that there stemmed a renewed, strong commitment to theory, viewed as one, and sometimes the only possible, form of praxis. Theory—and in particular philosophy, that, contra the prevailing positivist attitude, was identified as the privileged tool for the self-critique of reason and the necessary foundation for any serious theoretical sociology—was thus entrusted with a double task: first, to comprehensively confront the problem of dominion, and then to nurture the seeds of a noncompliant attitude hinging on the exercise of an "immanent critique" able to detect the societal contradictions offering actual possibilities for emancipatory change.[44] And when, in a later period, the anti-authoritarian rebellion of the students and the Third World's liberation movements opened the door to new hopes, the subsequent elaborations of the School were in a position to give a more constructive twist both to such attitude and to the overall emancipation program of Critical theory. In particular, it fell to Herbert Marcuse to raise the possibility of an opposition to the contemporary socio-economic order, and to offer actual instructions for a transformative praxis—instructions focused on the need for a cultural struggle giving priority to the subjective factor and operationalizing it in terms of the oppositional consciousness of a militant minority, capable to offer a much-needed source of radical imagination.[45]

Imagination seems instead to play, as has been suggested,[46] a lesser role within the second, more recent outlook one might draw political inspiration from, that is, Bourdieu's Reflexive sociology—a perspective whose Nietzschean emphasis on the intrinsic social conflict over power and prestige bars it from fostering palingenetic hopes.[47]

Reflexive sociology—a theoretical approach in which reflexivity plays the critical role of including within the field of scrutiny *both* the analyzed objects *and* the analyzing subjects—has produced a comprehensive anthropological view according to which, whatever its ideological or economic configuration, any society is the structural locus of struggles for power and status that can be predicated on something even more general than instrumental reason, a sort of "will to power."[48] Articulated on the basis of major theoretical analyses but also of close confrontations with specific phenomena like the Algerian decolonization or the events of May 1968, such a theory highlights the central role of competition both at the level of the division of the work of domination—among the sexes, the classes, and the various "fields," or structured systems of social positions, making up the social space—and at the level of each specific field, where agents dispute over particular advantages and privileges. Given that the winners in the social competition impose their power on the losers, creating dominant groups with enduring interests that favor the reproduction of their advantages,[49] the tendency to hierarchization and oppression turns out to be congenital to the entire social universe.[50]

But domination is only partially based on direct violence and economic power. Within the context of what has been seen as an extension of Marx's political economy from economic to symbolic goods,[51] Bourdieu argues that, if societies often work smoothly, with forms of coordination and cooperation within hierarchical and oppressive contexts, it is because they embody a principle of continuity which leads individuals to take the world for granted. Such a principle is the habitus,[52] a sort of socially constituted instinct operating with automatic reliability, a system of "structured and structuring structures" that are embedded in every member of the group and that crucially engage the basic principles expressing the doxa, or what goes without saying because it is both subjectively internalized and inscribed in the objective.[53]

All this clearly has important implications pointing toward the Frankfurtian outlook. For it means that hierarchy and oppression are strongly dependent on broadly cultural factors. And if the process whereby the social order extorts practical acceptance of its existing hierarchies is a form of symbolic violence—an imposition of *systems of meaning* that legitimize structures of inequality[54]—conversely, the dominated have the possibility to react through culture, by questioning the prevailing grids of classification and interpretation. Theory can thus be seen as *both* subordinate to the given *and* capable of transcending it,[55] especially at the junctures when external causal series introduce history—which mechanisms of reproduction tend to resist—into societies and fields.[56]

Thanks to culture, then, opposition to power, though difficult, is not impossible. And a philosophically problematized sociological analysis, interested in the demystification of the status quo, can provide instruments of defense against symbolic domination, thereby offering political tools to movements informed, if not by millenarian hopes, at least by realistic utopias.[57] In fact, Bourdieu himself, especially in the face of the recent expansion of global finance capital, directly put his conceptual arsenal at the service of the struggle against social injustice.[58]

All in all, then, despite the differences—not least the fact that, even though the notion of habitus as a sort of social instinct objectively erodes the traditional human/animal opposition,[59] Bourdieu's analyses never include nonhuman exploitation—there is a salient convergence between Critical theory and Reflexive sociology. For both approaches essentially believe that theory can have a transgressive use, and hold that a form of philosophical sociology can play a therapeutic role against repression, be it "by imagining the unimaginable, or by disturbing seemingly natural distinctions."[60]

In other words, starting from the plausible assumption that, if the capitalist mode of production is the structural element in the current exploitation of other animals, humanism is the superstructural factor, what the two interpretations here briefly outlined tell us, in different ways and appealing to different explanations, is, first, that the superstructural factor is far more important than was traditionally thought; and, second, that at this level, unlike what happens at the structural level, where the disparity of forces is overwhelming, political dissenters are not unarmed. And this because theory, both in the "passive" sense of offering a lucid construal of reality and in the "active" sense of offering instruments for intervention, is a powerful resource accessible even to the politically disadvantaged.

At this point, it should be easy to perceive the affinity of such approaches with a political enterprise that—already hindered by the lack of a decisive backing from the dominated group—must confront an extreme unfolding of instrumental reason and a vast exercise of symbolic violence largely supported by a mobilization of "cultural" resources. In this sense, it might be argued that a movement which, whilst waiting for a dubious revolution, sticks to the regulative ideas offered by the soundest declension of universalizability and the best liberal utopias, while at the same time accepting a logic of suspicion toward reality's structures and institutions, might find in Critical theory and Reflexive sociology some relevant political clues liable to assist it with developing its guiding principles in the current social context.

CREATING FUTURE FREEDOM

What are, then, the specific strategic ideas one can tentatively draw from the theoretical stances in question? Here are some suggestions that might hopefully provide a basis for discussion.

In a Frankfurtian perspective, a first step might be to focus on sorting out society's internal dialectics, so as to identify tendencies that might lead beyond the existing state of affairs.[61] As Marcuse puts it: to gain success, attempts to end intolerable conditions must imply that the logic invoked in the movement of thought and action is that of the given conditions to be transcended.[62] In this light, a praxis infused with theory will try to take advantage of the existing suppressed contradictions by activating them and making them operate as catalysts of change,[63] as well as by using the available instruments to guide them in the desired political direction.

Two current projects focusing on the nonhuman great apes and on cetaceans can exemplify this process. Demanding basic human rights for the nonhuman great apes—traditional victims of capitalism's dynamic in the form of experimentation by pharmaceutical companies and industrial destruction of the forests of the planet—and granting them moral and legal personhood is now made possible by the insertion of the arguments recently produced by rational ethics into the contradictions of an ambivalent cultural attitude that is the apex of a long history of wavering before their similarity and relatedness to our species.[64] In this enterprise, one might also profitably build on Bourdieu's diagnosis of cultural fields, according to which they are generally marked by a tendency of new entrants to adopt heretic stances and to deploy strategies of subversion to overturn established hierarchies and gain status.[65] Thus, quite apart from capitalizing on an ever more sympathetic public opinion, the animal liberation movement might here try to benefit from the conflict internal to the relevant scientific domains—from primatology to anthropology to interpecies communication and comparative psychology—where it has become rewarding for many authors to complete the destruction of the once dominating behaviorist paradigm, and its replacement by ever more subtle analyses of animal cognitive and emotional life. Actually, all this work is already starting to erode the paradigm, so much so that many countries have forbidden experimentation on our closest living relatives, and some pioneering parliamentary resolutions in favor of granting legal rights to the nonhuman great apes have been approved.[66]

In the case of cetaceans, on the other hand, an important role is being played by the "benign" side of the globalization of international law. For, with its gradual but steady proscription of massacres guided by industrial profitability, the evolution of the *opinio juris* of nations is adumbrating cetaceans' entitlement to life, thus opening the way for political pressure in favor of the articulation of the corresponding legal right.[67] As with the nonhuman great apes, such a pressure can benefit from the growing stress on the complexity of cetacean minds and cultures in the relevant scientific disciplines. Furthermore, drawing on Reflexive sociology's elucidation of the mechanism by which political events, in the presence of diverse fields pregnant with their own determinants, may be potentiated by a coincidence of different crises, so that mobilization can capitalize on their synchronization, the enterprise might try to avail itself of the energies liberated by the explosion of the environmental crisis, which fuels oppositional discourses favoring a strong conservationist stance.[68] An evidence of the progresses thus being made is the almost universal sympathy for the "whale wars" courageously waged against the fleets of noncompliant nations such as Norway and Japan.[69] A further sign is the decision of India's Ministry of Environment to forbid the keeping of dolphins in captivity, accompanied by a reference to dolphins as "nonhuman persons."[70]

In both cases—the one concerning the nonhuman great apes as well as the one concerning cetaceans[71]—an extension of basic equality could not only undermine the symbolic value of the species barrier, dispelling the social magic that produces discontinuity out of continuity[72] and creating a mechanism for further, wider extensions, but might also start to apply to the nonhuman realm that curbing of unruly economic exploitation that leftist political thinkers invoke only for the human domain.

Apart from favoring alliance building by catalyzing, rather than begging for, the support of differently minded but synergetic groups, these first experiments with transcending the given societal conditions could also offer, *hic et nunc*, a glimpse of what a possible Zoopolis might look like. For from the reality of present marine sanctuaries could gradually arise a sort of sovereignty of cetaceans over the oceans they inhabit; and, in the case of the nonhuman great apes who, after a long captivity, must necessarily keep living among us, a first experiment in citizenship might be done, as is already foreshadowed by the ongoing initiatives directly engaging the legal domain with the aim of changing the juridical status of currently imprisoned individuals by obtaining for them legal personhood.[73] Actually, the theme of citizenship could play a role in another locus of contradiction that has been explored ever since the movement's inception and is being reconsidered now with different theoretical tools,[74] that is, the realm of companion animals, who are in many countries variously protected and sometimes even considered in courts as family members, with the consequence that much pressure is exerted on their current status of property. With respect to such a realm, there is, despite the fact that the public intellectual debate is just beginning,[75] the advantage of the possible support of a wide, extremely motivated, and easily mobilizable section of public opinion.

As Critical theorists emphasized, in all these enterprises, radicalism has much to gain from "legitimate" political activity as well as from a reflexive use of the existing legal frameworks and of the democratic resources which are the outcome of past struggles.[76] Focused projects of this kind, however, cannot remain isolated, as they are only the tips of the iceberg of the global struggle for the eradication of all animal exploitation. And, to fight such a struggle, where a direct radical intervention is often impossible, a further concept on which the Frankfurtians elaborated might possibly offer a cue—the idea of a long march through the institutions.[77] The sense of this policy is clear: if society cannot be directly changed, it should be modified in a mediated way by influencing its ganglions—that is, by working against established institutions while working in them. Manifestly central to the strategy is a clever use of theory: the altering of cultural hegemony through institutional entryism—the penetration into social casemates like the publishing system, magistracy, and especially universities, key institutions in the training of counter-cadres—as well as through concerted efforts to create counter-institutions.

A process of this kind is already under way, and deepening and expanding it should not be too problematic for a social entity whose cornerstone is academic, and whose scholars might, along Bourdieusian lines, build up a "collective intellectual" capable of placing multiple competences in the service of the cause.[78] As mentioned, political subversion presupposes cognitive subversion, and in particular the unmasking of the process of euphemization through which the dominant make violence unrecognizable by conferring on it the appearances of the natural.[79] And, from the very efforts of the defenders of the status quo, even when they hold all the keys to power, to produce an incessant propaganda flooding the public with reinstatements of the doxa,[80] it can be inferred how vulnerable oppressive relations are to a heretical discourse that can arouse formerly censored feelings and tendencies, conferring the objectivity of public theory to a perspective on the world previously relegated to the state of a tacit or confused experience.[81]

The critical work of feminist theorists has, at least in some regions of social space, succeeded in breaking the circle of the generalized mutual reinforcement of prejudices. Analogously, the theoretical dismantling of the justifications of the paradigm of human superiority, resolutely conducted in defiance of all the forms of compulsory discursivity that govern the conditions under which political claims on behalf of animals can be made—above all, the requirement of a preliminary kowtowing to the humanist ideology[82]—and based not only on direct argumentation but also on the deconstruction and reconstruction of doxic conventions,[83] might seriously challenge and influence, through a top-down process, the public acceptance of such practices as flesh-eating and using nonhumans as instruments for work or research.

Of course, to be effective, as well as to help ensure that militant mobilization might produce effective political campaigns, this enterprise should be accompanied by another fundamental undertaking—the carrying on, through processes of discursive identification of the social actors, the symbolic work that is the precursor of mobilization itself.[84] Such a work, that was clearly instrumental to the social construction of classes, and that centers around establishing a group as a relevant cultural and social force, at once expressed and constituted by theoretical views, political stances, constituents and representatives, is even more relevant for a movement which, inasmuch as its precursors tended to be seen as conservative crutches of the established order or even as the model of those marginal organizations that nobody could take seriously,[85] is in definite need to assert itself as a radical political endeavor pursuing institutional changes.

Finally, a related point concerning unity.[86] It is often regretted that the animal liberation movement is split by personal and political antagonism. This may be true, but it's true of any other collective social agent. The problem is rather that, disregarding its capacity to organize itself, the movement has not generated a global, coordinated network, able to establish connections among the various centers into which it is divided. If developed, such a network not only could make initiatives more conspicuous by bringing into existence in an instituted form a collective agent hitherto present only in a potential state, but could also play the crucial role of providing a locus for the clash between opposing views. For it is mainly in such clashes that political strategies can be debated and tested.

The Presence of Animals

In conclusion, let us return to the stigmatized beings. We suggested that one of the difficulties the animal liberation movement must face is the lack of a determinant support from the dominated group it defends. This does not mean, however, that animals are devoid of social agency, or that their autonomous acts can be simply framed as abstract, ideal gestures or as generic manifestations of an antagonism issuing from "nature's resistance."[87]

A history from the point of view of nonhuman beings is only now starting to be written. And this not in the sense of presenting historical facts in a light stressing the role played by the exploitation of animals in determining major military or political events,[88] but in the literal sense of casting animals as the main characters of events. Previously, the rare and indirect descriptions of the subjective responses of nonhumans to the procedures and practices affecting their lives were relegated to the sphere of the anecdotal. On the one hand, this was in consonance with the universal tendency to dismiss anything other than human. On the other, it was a consequence of the doxic philosophical view that history is a superior realm to which only humans have access, while nonhumans, both individually and collectively, although exposed to externally induced changes, remain subjectively in the grip of an eternal present.

The path to the new approach was smoothed by the discovery that autonomous animal communities do have a history, to whose course individual nonhumans can make a difference.[89] The gestalt shift subsequently involved the more intricate question of the history of animals enmeshed in human societies. Thanks to a multidisciplinary approach ranging from physiology to ethology to cognitive studies, "another version of history"[90] brought to the forefront the animal perspective on some major turns in the exploitation process, thus making it possible to start identifying their actual impact. And to comprehend the consequences of such turns from the subjective standpoint of nonhumans means not only to grasp the new forms of pressure brought to bear on their natural characteristics, above and beyond what a basic role shift test could suggest,[91] but also, even more relevantly, to figure out how the concerned individuals reacted.

In general, reactions to oppression can be clustered around the two poles of acquiescence and resistance, each of which can involve different balances of passivity and activity. Clearly, given the overwhelming disparity in power, most nonhumans, usually anonymous members of an exploited mass, could not but unconditionally acquiesce. This means that their welfare and their lives were totally crushed by human interventions. In some cases, however, when it was possible to develop an individual relationship with the involved "keepers," acquiescence took the more active form of adaptation through the introjection of human desiderata. And the great sophistication displayed in this process—suffice it to think of the flexibility and subtlety of dogs[92]—should invite an extension beyond the human realm of that section of subaltern studies which focuses on the internalization of dominant values by the oppressed.

Against all odds, however, the subjugation processes also elicited manifestations of animal resistance.[93] In other words, they evoked that relevant phenomenon of *struggle* whose presence can be seen as the main hallmark of history.[94] In this case too, the less active facet tends to prevail: essentially, animals take to flight, directly or after an act of rebellion,[95] even though one also finds impressive forms of strike, such as when some of the cows confined in the first centralized dairy facilities managed to retain their milk in order to save it for their calves at the price of painful engorgements. At the opposite extreme, on the other hand, resistance can take the shape of retaliatory violence.

Instances of retaliatory violence by nonhumans have never been lacking—suffice it to recall the public electrocution or hanging [*sic!*] of elephants who had killed their "trainers."[96] In recent periods, though, there has been a steady growth in reports of episodes of animal vengeance. Tigers clear the walls of their prisons to assault individuals who tormented them, searching for them, and ripping them to pieces[97]; orcas trapped in small concrete tanks drown their instructors by pinning them to the bottom of the tanks until they die[98]; exasperated elephants attack all over Africa and India the forest settlers who are causing the collapse of their societies, rampaging through villages and terrorizing and killing their inhabitants[99]; and traumatized bulls develop tactics to kill matadors or aficionados by repeatedly goring them and then throwing them in the air.[100]

According to the British moralists, the feeling of revenge that underlies retaliatory violence is a component of the sense of justice, and has long been one of the chief restraints to crime.[101] Thus, when, as is the case with nonhumans in our societies, there is no hope for public justice, and there are no further restraints to crime, retaliatory violence appears to fill a vacuum. True, isolated gestures cannot concretely affect the power of the dominant. But a desperate act may for a brief moment expose the faces of brutal suppression, and, arousing the conscience of the neutrals,[102] can illuminate the struggle and draw new lifeblood into it. Moreover, in its immediateness and noninstrumentality, the retaliatory violence of oppressed animals is evocative of what the most heterodox among the Frankfurtians, Walter Benjamin, once characterized as divine violence—a phenomenon which, in contrast with the forms of legal or mythical violence whose principle is power, is neither law-making nor law-preserving,[103] and, as a consequence, does not herald any further perpetrations of violence. In the face of all this, contra any reductionist view, one can argue that, through the symbolic implications of their acts, revengeful animals can indeed make a significant contribution to the animal liberation movement's conflict with the hegemonic order.

NOTES

1. Peter Singer, "All animals are equal," *Philosophical Exchange*, 1/5 (1974): 103–116.
2. Fernando Pessoa, "The Anarchist Banker," in *The Anarchist Banker and Other Portuguese Stories*, ed. Eugenio Lisboa (Manchester: Carcanet Press Ltd., 1997), available at www.csend.org/component/docman/doc.../ 284-20101009-fernandopessoa.
3. On humanism, see, for example, Paola Cavalieri, "Consequences of Humanism, or, Advocating What?," in *Species Matters. Humane Advocacy and Cultural Theory*, ed. Marianne DeKoven and Michael Lundblad (New York: Columbia University Press, 2012), 49–74. On commodification see, for example, Gregory Smulewicz-Zucker, "The Problem with Commodifying Animals," in *Strangers to Nature: Animal Lives and Human Ethics*, ed. Gregory Smulewicz-Zucker (Lanham, MD: Lexington Books, 2012), 157–

174; and Nicole Shukin, *Animal Capital: Rendering Life in Biopolitical Times* (Minneapolis: University of Minnesota Press, 2009).

4. Hannah Arendt, *The Origins of Totalitarianism* (New York: Harcourt, Brace, 1973), 440.

5. For a survey and critique, see Cary Wolfe, *Before the Law. Humans and Other Animals in a Biopolitical Frame* (Chicago: The University of Chicago Press, 2013).

6. See Slavoj Žižek, "On Alain Badiou and Logiques des mondes," http:// www.lacan.com/zizbadman.htm.

7. See, for example, Jiří Navrátil, "Dialectics of Revolt. Changes of the Social Movement Concept in the Political Theory of the Frankfurt School" (in Czech), *Czech Journal of Political Sciences* 4 (2007): 321–335.

8. For the former pole, see, for example, David Macaulay, "Political Animals: A Study of the Emerging Animal Rights Movement in the United States," *Between the Species*, from 3/2 (1987) to 4/2 (1988), and Lawrence Finsen and Susan Finsen, *The Animal Rights Movement in America* (New York: Twayne Publishers, 1994). For the latter, see, for example, Gary Steiner, *Animals and the Limits of Postmodernism* (New York: Columbia University Press, 2013), chap. 5.

9. See Lee Hall, "An Interview with Professor Gary L. Francione on the State of the U.S. Animal Rights Movement," *Friends of Animals*, Summer 2002, http://friendsofanimals.org/programs/animal-rights/interview-with-gary-francione.html.

10. See Tom Regan in "A Discussion with Tom Regan," *Ahimsa* Oct/Dec 1987.

11. See, for example, Dale Jamieson, "Against Zoos," in *In Defense of Animals*, ed. Peter Singer (Oxford: Blackwell, 1985), 108–117; or Lewis Regenstein, "Animal Rights, Endangered Species and Human Survival," ibid., 118–132.

12. Steve Best, "Rethinking Revolution: Animal Liberation, Human Liberation, and the Future of the Left," *The International Journal of Inclusive Democracy* 2/3 (2006), http://www.inclusivedemocracy.org/journal/vol2/vol2_no3_Best_rethinking_revolution.htm.

13. Matthew Calarco, "Review of *In Defense of Animals: The Second Wave*," *Journal for Critical Animal Studies* 5 (2007): 82–87.

14. See Judith Butler, *Excitable Speech: A Politics of the Performative* (New York: Routledge, 1997), 90.

15. See "Hegemony and Socialism: An Interview with Chantal Mouffe and Ernesto Laclau," *Palinurus. Engaging Political Philosophy* 14 (April 2007), available at anselmocarranco.tripod.com/.

16. Slavoj Žižek, "The Prospects of Radical Politics Today," *International Journal of Baudrillard Studies*, 5/1 (2008), http://www2.ubishops.ca/

baudrillardstudies/vol5_1/v5-1-article3-zizek.html. See on this Paola Cavalieri and Peter Singer, "Reply to Žižek. On Žižek and Animals," *International Journal of Baudrillard Studies*, 6/1 (2009), http://www2. ubishops.ca/baudrillardstudies/vol-6_1/v6-1-Singer-cavalieri.html.

17. See Ted Benton and Simon Redfearn, "The Politics of Animal Rights— Where is the Left?," *New Left Review* 215 (1996): 43–58.

18. For an interesting analysis of the decline of the welfare state see Wolfgang Streeck, *Buying Time: The Delayed Crisis of Democratic Capitalism* (London: Verso, 2014). In the book, Streeck explores the problems of contemporary capitalism through a revisitation of the Frankfurt School crisis theories.

19. See, for example, Ted Benton, *Natural Relations: Ecology, Animal Rights and Social Justice* (London: Verso, 1993); Renzo Llorente, "Reflections on the Prospects of a Non-Speciesist Marxism," in *Critical Theory and Animal Liberation*, ed. John Sanbonmatsu (Lanham, MD: Rowman and Littlefield, 2011), 121–135; Katherine Perlo, "Marxism and the Underdog," *Society & Animals* 10/3 (2002), 303–318; and David A. Nibert, *Animal Rights- Human Rights: Entanglements of Oppression and Liberation* (Lanham, MD: Rowman & Littlefield, 2002).

20. See, for example, Matthew Calarco and Peter Atterton, eds, *Animal Philosophy: Essential Readings in Continental Thought* (New York: Continuum, 2004).

21. For more detailed examples, see Paola Cavalieri, "Consequences of Humanism," 50.

22. On the pitfalls of such approaches, see, for example, Harlan B. Miller, "Distracting Difficulties," in Paola Cavalieri, *The Death of the Animal. A Dialogue*, with commentaries by M. Calarco, J.M. Coetzee, H.B. Miller, and C. Wolfe (New York: Columbia University Press, 2009), 111.

23. See Sue Donaldson and Will Kymlicka, *Zoopolis. A Political Theory of Animal Rights* (Oxford: Oxford University Press, 2011).

24. On this, see Étienne Balibar, *Cittadinanza* (Torino: Bollati Boringhieri, 2012), 80.

25. Arendt, *The Origins of Totalitarianism*, Chap. 9.

26. Ibid., 296.

27. Balibar, *Cittadinanza*, 14, 38.

28. See Max Weber, "Capitalism and Rural Society in Germany," in *From Max Weber: Essays in Sociology*, ed. H.H. Gerth and C. Wright Mills (Oxford: Oxford University Press, 1958), 365.

29. For the distinction between micro-insurgency and revolution, see Bob Stone, "Why Marxism Isn't Dead (Because Capitalism Isn't Dead)," *Paideia Archiv* 11 (Social Philosophy), 2001, www.bu.edu/wcp/Papers/Soci/ SociSton.htm.

30. Jacques Derrida, "L'esprit de la révolution," in Jacques Derrida and Elisabeth Roudinesco, *De quoi demain... Dialogue* (Paris, Fayard/Galilée, 2001), 138.

31. See Michael Hardt and Antonio Negri, *Multitude* (New York: The Penguin Press, 2004), 20; and, more recently, M. Hardt and A. Negri, *Declaration* (New York: Argo-Navis Author Services, 2012), esp. "Introduction."

32. See Alain Badiou, *The Rebirth of History* (London/New York: Verso, 2012), 56 and passim. For the reference to theology see Daniel Bensaïd, "Alain Badiou and the Miracle of the Event," in *Think Again: Alain Badiou and the Future of Philosophy*, ed. Peter Hallward (London/New York: Continuum, 2004), 94–105.

33. See Slavoj Žižek, "Repeating Lenin," http://www.lacan.com/replenin.htm.

34. It is also worth noting not only that such insurgencies are mostly the deed of untamed young people who are wholly unaware of current Leftist theorizations, but also that, as many of their developments show, they often turn out to be fueled by a split within the ruling class, or even by external interests.

35. The phrase is Deutscher's: see Isaac Deutscher, *Heretics and Renegades and Other Essays* (London: Jonathan Cape Ltd., 1969).

36. For the distinction, see Walter Benjamin, "Critique of Violence," in W. Benjamin, *Selected Writings, Vol I* (Cambridge, MA: Belknap Press, 1999), 277–300.

37. Max Horkheimer and Theodor W. Adorno, *Dialectic of Enlightenment* (Palo Alto, CA: Stanford University Press, 2002), 65 ff.

38. On this, see Göran Therborn, "The Frankfurt School," *New Left Review* 63 (1970).

39. See, for example, Steve Best, "Rethinking Revolution"; and Christina Gerhardt, "Thinking With: Animals in Schopenhauer, Horkheimer and Adorno," in Sanbonmatsu, *Critical Theory and Animal Liberation*, 137–146.

40. For Horkheimer and Adorno on the animal question, see Horkheimer and Adorno, *Dialectic of Enlightenment*, "Man and Beast," 203 ff. and passim. On this, see Paola Cavalieri, "Do We Need Continental Philosophy? Nonhumans, Ethics, and the Complexity of Reality," in David L. Clark, ed., "Animals . . . In Theory: Nine Inquiries in Human and Nonhuman Life," Special Issue, *The New Centennial Review* 11/2 (2011): 83–113.

41. See Max Horkheimer, "Wolkenkratzer," in his *Dämmerung. Notizen in Deutschland* (Frankfurt: Fischer, 1974), first published in Zurich in 1934 under the alias of Heinrich Regius, quoted in Renate Brucker, "Animal Rights and Human Progress" (paper presented at the conference Animals in History, Literaturhaus, Cologne, Germany, May 18–21, 2005).

42. Horkheimer and Adorno, *Dialectic of Enlightenment*, 43.

43. See Tito Perlini, "Dall'utopia alla teoria critica. Benjamin e la Scuola di Francoforte," *Comunità* 159/160 (1969).

44. On immanent critique, see, for example, Theodor Adorno, *Negative Dialectics* (New York: Continuum, 1973), 181 ff.; and Max Horkheimer, "Traditional and Critical Theory," in id., *Critical Theory: Selected Essays* (New York: Continuum, 1975), 215. On the role of immanent critique within Critical theory see Robert J. Antonio, "Immanent Critique as the Core of Critical Theory," *The British Journal of Sociology* 32/3 (1981): 330–345.

45. See Jiří Navrátil, "Dialectics of Revolt."

46. See Nedim Karakayali, "Reading Bourdieu with Adorno: The Limits of Critical Theory and Reflexive Sociology," *Sociology* 38/2 (2004): 351–368.

47. See Mitchell Aboulafia, "A (neo)American in Paris. Bourdieu, Mead and Pragmatism," in *Bourdieu: A Critical Reader*, ed. Richard Shusterman (Oxford, Blackwell, 1999), 157; and more generally on Bourdieu and Nietzsche, see Jen Webb, Tony Schirato, and Jeoff Danaher, *Understanding Bourdieu* (London: Sage Publications, 2005), 14 ff.

48. The suggestion comes from Webb, Schirato, and Danaher, who make a direct reference to Friedrich Nietzsche. See ibid., 14, 20.

49. See Bridget Fowler, "Pierre Bourdieu: Unorthodox Marxist?," in *The Legacy of Pierre Bourdieu*, ed. Simon Susen and Bryan S. Turner (London: Anthem Press, 2011), 33–58.

50. See Pierre Bourdieu and Loïc Wacquant, *Invitation to Reflexive Sociology* (Cambridge: Polity Press, 1992), 212.

51. See Michel Burawoy, *Conversations with Pierre Bourdieu, IV. Colonialism and Revolution: Fanon Meets Bourdieu*, http://burawoy.berkeley.edu/Bourdieu/5.Fanon.pdf.

52. On habitus, see, for example, Pierre Bourdieu, *The Logic of Practice* (Cambridge: Polity Press, 1990), 53, 104, 161; id., *Outline of a Theory of Practice* (Cambridge: Cambridge University Press, 1977), 17–18, 21, 79; id., *Practical Reason. On the Theory of Action* (Stanford, CA: Stanford University Press, 1998), 54. Bourdieu has given new depth to a notion that runs through all Western philosophy—from Aristotle to Aquinas to Husserl—to refer to a system of dispositions existing in a virtual state, and manifesting themselves in specific conditions.

53. On doxa, see, for example, Bourdieu, *Outline of a Theory of Practice*, 159–171, and Pierre Bourdieu, *Distinction. A Social Critique of the Judgement of Taste* (London: Routledge, 2010), 473.

54. On symbolic violence, see, for example, Pierre Bourdieu, *Masculine Domination* (Stanford: Stanford University Press, 2001), 34 ff.; see also Loïc Wacquant, "Pierre Bourdieu," in *Key Sociological Thinkers*, ed. Rob Stones (second edition, London and New York: Palgrave Macmillan, 2008), 261–277.

55. See Karakayali, "Reading Bourdieu with Adorno," 360 ff.

56. See Pierre Bourdieu, *Homo Academicus* (Cambridge: Polity Press, 1988), 33.

57. On the instruments of defense offered by Reflexive sociology, see Pierre Bourdieu, "For a Scholarship with Commitment" (keynote address delivered to the Modern Language Association Meeting, Chicago, December 27–30, 1999), now in *Sociology is a Martial Art. Political Writings by Pierre Bourdieu*, ed. Gisele Sapiro (New York: The New Press, 2010), 178–185.

58. For Bourdieu's fight against injustice, see Karakayali, "Reading Bourdieu with Adorno"; more generally, on Bourdieu's trajectory see his short portrait in Wacquant, "Pierre Bourdieu."

59. For Bourdieu's implicit overriding of the cognitive opposition between humans and animals, see, for example, Paola Cavalieri, "Declaring Whales' Rights," in "Cetacean Nations," Special issue, *Tamkang Review*, 42/2 (2012): 124 ff.

60. Karakayali, "Reading Bourdieu with Adorno," 364.

61. See Herbert Marcuse, *An Essay on Liberation* (Boston: Beacon Press, 1969), 3.

62. Herbert Marcuse, *One-dimensional Man* (Boston: Beacon Press, 1991), 223.

63. See Herbert Marcuse, "Liberation from the Affluent Society" (lecture held at the conference "Dialectics of Liberation," London, July 15–30, 1967), published in *The Dialectics of Liberation*, ed. David Cooper (Harmondsworth/Baltimore: Penguin, 1968), 189, available at http://www.marcuse.org/herbert/pubs/60spubs/67dialecticlib/67LibFromAfflSociety.htm. Of course, contradictions are not static, but can vary as the situation varies.

64. See, for example, Paola Cavalieri and Peter Singer, eds, *The Great Ape Project. Equality beyond Humanity* (London: Fourth Estate, 1993). See also Robert Goodin, Carole Pateman and Roy Pateman, "Simian Sovereignty," *Political Theory* 25/6 (1997): 821–849; Raymond Corbey, *The Metaphysics of Apes: Negotiating the animal-human Boundary* (Cambridge and New York: Cambridge University Press, 2005), and Paola Cavalieri, "The Meaning of the Great Ape Project," *Politics and Animals* 1 (2015). For the social level, see, for example, New Zealand at http://www.animallaw.info/journals/jo_pdf/lralvol_7p35.pdf, Spain at http://www.guardian.co.uk/world/2008/jun/26/humanrights.animalwelfare and Germany at http://www.giordano-bruno-stiftung.de/sites/default/files/download/3_petition51830_chronologie.pdf.

65. See Pierre Bourdieu, *The Rules of Art* (Stanford, CA: Stanford University Press, 1996), 239 ff.; see also Pierre Bourdieu, "Quelques propriétés des champs," in his *Questions de sociologie* (Paris: Éditions de Minuit, 1980), 113–120.

66. See Peter Singer, "Of Great Apes and Men," *The Guardian*, July 18, 2008.

67. See Anthony D'Amato and Sudhir K. Chopra, "Whales: Their Emerging Right to Life," *American Journal of International Law* 85/1 (1991), 21–62; Cavalieri, "Declaring Whales' Rights"; Thomas I. White, *In Defense of Dolphins: The New Moral Frontier* (Oxford: Blackwell, 2007).

68. See Pierre Bourdieu, *Homo Academicus*, 161, 173, 180 ff. For an interesting application of this grid, see Marcos Ancelovici, "Esquisse d'une théorie de la contestation: Bourdieu et le modèle du processus politique," *Sociologie et sociétés* 41/2 (2009), 39–61.

69. See Paul Watson, *Sea Shepherd* (New York: W. W. Norton and Company, 1980); David Day, *The Whale War* (Vancouver: Douglas and McIntyre, 1987); and http://www.seashepherd.org/whales.

70. On the statement of India's Ministry of Environment, see "India Bans Captive Dolphin Shows as 'Morally Unacceptable,'" Latest News, RSS, Wildlife, May 20, 2013, http://ens-newswire.com/2013/05/20/india-bans-captive-dolphin-shows-as-morally-unacceptable/.

71. Weren't it for an aborted beginning, from which it should hopefully recover, one might also mention a parallel initiative concerning elephants; see the moderate conclusions of Michael J. Glennon's "Has International Law Failed the Elephant?," *The American Journal of International Law* 84/1 (1990), or the advances and the withdrawals in the text of the Elephant Charter, http://www.theelephantcharter.info/. But see at least the recent defense of elephant sovereignty in Paola Cavalieri, "Elephants under fire," *New Scientist* 3071 (30 April 2016).

72. See Pierre Bourdieu, *Language and Symbolic Power* (Cambridge: Polity Press, 1991), 120.

73. See, for example, Steve Wise's Nonhuman Rights Project, http://www.nonhumanrightsproject.org/steve-wise/, and, in particular, the Project's attempt to obtain the status of "legal person" for the chimpanzee Tommy, http://www.nytimes.com/2013/12/03/science/rights-group-sues-to-have-chimp-recognized-as-legal-person.html; see also Martin Balluch's application for guardianship for the chimpanzee Matthew Hiasl Pan, a description of which can be found in Nichola Donovan, "Challenging the 'art of disinformation' in Australian animal law" (paper delivered at the Australian Law Librarians' Association Annual Conference, Perth, Australia, Sept. 19, 2008), http://www.alla.asn.au/conference/2008/docs/donovan.doc. For an ethnographic survey of the specificities of operating within the legal terrain see Ciméa Barbato Bevilaqua, "Chimpanzees in court: what difference does it make?," in *Law and the Question of the Animal*, ed. Yoriko Otomo and Ed Mussawir (London: Routledge, 2013), 71–88.

74. See Joyce S. Tischler, "Rights for Non-human Animals: A Guardianship Model for Dogs and Cats," *San Diego L. Rev.* 14 (1977): 484–506 and Sue

Donaldson and Will Kymlicka, "Citizen Canine: Agency for Domesticated Animals" (paper presented at "Domesticity and Beyond: Living and Working with Animals," Queen's University, Sept. 29–30, 2012).

75. On dogs, see, for example, Gregory Berns, "Dogs are people, too," *The New York Times*, October 6, 2013, available at http://www.nytimes.com/2013/10/06/opinion/sunday/dogs-are-people-too.html?_r=0. On cats see, for example, Frank Bruni, "According Animals Dignity," *The New York Times*, January 13, 2014, http://www.nytimes.com/2014/01/14/opinion/bruni-according-animals-dignity.html?hpw&rref=opinion.

76. See Herbert Marcuse, *Counterrevolution and Revolt* (Boston: Beacon Press, 1972), 56.

77. Ibid., 55. Here, Marcuse openly borrows from the slogan that Rudi Dutschke, the German student-movement leader, launched in 1967.

78. See Pierre Bourdieu, "The Intellectual is not Ethically Neutral," in Sapiro, *Sociology is a Martial Art*, 259.

79. See Pierre Bourdieu, "For a Real Mobilization of Organized Forces," in Sapiro, *Sociology is a Martial Art*, 248.

80. See, for example, Bourdieu, *Practical Reason*, 19, 114 and passim.

81. See Bourdieu, *Language and Symbolic Power*, 127 ff.

82. On compulsory discursivity, see, for example, J. Butler, "Ruled Out: Vocabularies of the Censor," in *Censorship and Silencing: Practices of Cultural Regulation*, ed. Robert Post (Los Angeles, CA: The Getty Research Institute, 1998), 254.

83. The prime example is of course the introduction of the terms *nonhumans* and *other animals*.

84. See, for example, Pierre Bourdieu, "What Makes a Social Class? On the Theoretical and Practical Existence of Groups," *Berkeley Journal of Sociology* 32 (1987), 13 ff. For a synthesis of Bourdieu's position on the officialization of groups, see Elliot B. Weininger, "Pierre Bourdieu on Social Class and Symbolic Violence," in *Approaches to Class Analysis*, ed. E. O. Wright (Cambridge: Cambridge University Press, 2005), 148–151 and 161–162.

85. For the reference to crutches, see Karl Marx and Friedrich Engels, *Manifesto of the communist party* (Radford, VA: Wilder Publications, 2007), 39; for the reference to marginality, see Arendt, *The Origins of Totalitarianism*, 292. By the way, apropos of the enduring discrepancy between the movement and what persists of the traditional forms of animal defense, there is a further suggestion that can be drawn from the notion of a "long march." For, when they do not directly collide with the movement's agenda, the requests advanced by conservative groups have the effect of raising the issue, especially in countries traditionally impervious to it, and can even succeed in affecting the institutional framework. This, in turn, could pave the

way for more radical requests as, due to bureaucracy's propensity to perpetuate itself, once agencies focusing on animal protection are established, they will be there to stay and, in order to last, will tend to play a progressive role. In this light, it might be suggested that an ideologically confident animal liberation movement could consider setting aside a merely principled approach in favor of the more strategic idea of a further form of entryism, through which the work of conservative groups can be overseen and perhaps also influenced.

86. On this, see, for example, Pierre Bourdieu, "Against the Policy of Depoliticization," in Sapiro, *Sociology is a Martial Art*, 203, and Herbert Marcuse, *Counterrevolution and Revolt*, 36–37.

87. See, for example, Shukin, *Animal Capital*, 86.

88. On this, see David A. Nibert, *Animal Oppression and Human Violence. Domesecration, Capitalism and Global Conflict* (New York: Columbia University Press, 2013).

89. Individual nonhumans can make a difference through their ways of managing internal conflicts, their reactions to major events like epidemics, their attitudes toward, or role in, hostilities and warfare, and of course their ability to introduce innovations concerning technique or tactics. See, for example, for chimpanzees Jane Goodall, "Chimpanzees – Bridging the Gap," in Cavalieri and Singer, *The Great Ape Project*, 12; or for killer whales, Vanessa Victoria, "An Acoustic Measurement of the Leadership Role/s of the Matriarchs in the SRKW Population of the Salish Sea," Friday Harbor Labs, WA: Beam Reach Marine Science and Sustainability School, 2010, http://www.beamreach.org/wp-content/uploads/vanessa102-proposal.pdf; or for elephants Joyce Poole, "An Exploration of a Commonality between Ourselves and Elephants," in "Nonhuman Personhood," Special issue, *Etica & Animali* 9 (1998), 85–110.

90. See Éric Baratay, *Le point de vue animal. Une autre version de l'histoire* (Paris: Seuil, 2012). The subsequent considerations and examples are mainly borrowed from this book.

91. It means, for example, to understand what it was like for the horses lowered into the narrow passages of the mines not to be able, due to the position of their eyes, to detect the origin of the noises that terrified them: or for the naturally sociable cows, during the "dairy conversion," to be suddenly deprived of any relationships with their chosen fellows, and to be tied up beside disquieting strangers; or for the bulls thrown into the arena not to manage to grasp the situation due to their downward gaze, and to be forced to forget their timidity of herbivores to violently react out of confusion and terror. See Baratay, *Le point de vue animal*, passim.

92. Ibid., 298 ff.

93. For an analysis of nonhuman acts of resistance and of their impact on their treatment, see Jason Hribal, "Animals, Agency and Class: Writing the History of Animals from Below," *Human Ecology Review* 14/1 (2007).

94. On struggle and history, see Pierre Bourdieu, *Sociology in Question* (London: Sage Publications, 1993), 40.

95. Thus, in the XIX century, underfed and overworked dogs seek an autonomous life by creating urban canine gangs analogous to the contemporary urban children's gangs; and when the creation of "pure breeds" gains momentum, many of the selected dogs rebel and run away to attend to the partners they desire, with the result of being massacred, together with their progeny. On the other hand, when the production is intensified in the mines, horses can refuse to work, often unfastening the chains of the trucks, and even, once back on the surface, managing to free themselves; and, among the horses sent to the First World War front lines, where they would die in the thousands, some try not to climb on the wagons, or, once reached the barracks, bite or unseat their rider and flee away.

96. On the electrocution of the elephant Topsy see http://www.sheepsheadbites.com/2013/01/topsy-the-elephant-100-years-later/; on the hanging of the elephant Mary ("Murderous Mary") see http://www.ashvegas.com/the-day-they-hanged-mary-the-elephant. Other well-known cases concern persecuted sperm whales fighting back to defend their fellows in danger, and courageously staving and sinking whaling and non-whaling ships; see, for example, the report by the explorer J.N. Reynolds originally appeared in *The Knickerbocker, or New-York Monthly Magazine* 13/5 (1839), 377–392, available at http://mysite.du.edu/~ttyler/ploughboy/mochadick.htm.

97. See Jason Hribal, *Fear of the Animal Planet: The Hidden History of Animal Resistance* (Oakland, CA: CounterPunch and AK Press, 2010), 21 ff.

98. See http://www.heraldtribune.com/article/20100225/article/2251077

99. See http://www.livescience.com/14541-killer-elephants-india-attack.html

100. See http://www.guardian.co.uk/world/2011/aug/14/bull-kills-third-man-spain

101. See, for example, Henry Sidgwick, *The Methods of Ethics* (7th edition, Indianapolis: Hackett Publishing Company, 1981), Book IV, Chap. III, §2.

102. Marcuse, *Counterrevolution and Revolt*, 52.

103. See Benjamin, "Critique of Violence," 297–300. On this question see also Andrew Robinson, "Walter Benjamin: Critique of the State," *Ceasefire*, December 31, 2013, http://ceasefiremagazine.co.uk/walter-benjamin-critique-state/.

Reorienting Strategies for Animal Justice

Matthew Calarco

I have been reading Paola Cavalieri's writings with great interest and appreciation for some two decades. Her steadfast desire to develop philosophical and theoretical perspectives aimed at achieving genuine ethico-political transformations in human-animal interactions has repeatedly drawn me to her work; despite certain differences in our respective political and philosophical commitments, I feel fortunate to count her as an ally in the struggle for achieving animal justice. Beyond our shared commitment to animal liberation, I also respect Cavalieri's willingness to draw from a wide variety of theoretical and disciplinary frameworks in order to develop critical tools for animal defense. While our home discipline of philosophy has a tendency to encourage practitioners to stay in their respective philosophical camps, Cavalieri has boldly made use of leading trends in both analytic and Continental philosophy and in cutting-edge work in the cognitive and physical sciences. The kind of multi-perspectival and interdisciplinary work modeled by Cavalieri is precisely what is needed as

M. Calarco (✉)
Department of Philosophy, California State University, Humanities Building 311, Fullerton, CA 92834-6868, USA
e-mail: mcalarco@Exchange.FULLERTON.EDU

© The Author(s) 2016
P. Cavalieri (ed.), *Philosophy and the Politics of Animal Liberation*,
DOI 10.1057/978-1-137-52120-0_3

we think through the question of how we might reorient strategies[1] for animal justice.

The chapter from Cavalieri that forms the focus of the present volume is of particular interest inasmuch as it helps to elucidate the general strategies at work behind some of the specific philosophical arguments and political initiatives for which she is best known. Clarity concerning strategy is important here, as some of Cavalieri's readers have taken her emphasis on rights for great apes and cetaceans as an instance of a kind of hierarchical animal ethics and politics, wherein animals who are most similar to human beings somehow have more value and are more deserving of political standing. As she has been at pains to make clear in several writings, such perfectionism and sliding scales of value are anathema to her work.[2] Rather, the point is that rights for great apes and cetaceans will presumably be easier to secure, given the dominant discourse surrounding the requisite properties of rights holders and what we currently know about the biological and cognitive capacities of these particular animals. Still, one could be left wondering where the larger stakes of such a project lie. How will rights for one or two species who are extremely similar to human beings in ethically and politically relevant ways be of benefit to other animal species that are less like us? Doesn't limiting our focus on securing rights for a handful of animals risk leading to the neglect of the many?

Cavalieri's chapter in this volume sheds light on these broader strategic stakes using theoretical and political traditions, namely, Frankfurt School-style Critical Theory and Bourdieu's Reflexive Sociology, that are quite different from those typically employed by analytic philosophers. The strategy most often employed by analytic animal ethicists is the same one that has dominated in applied ethics for many years now: defend what is taken to be the most cogent ethical theory, apply it to animals, and simply let the political implications of that approach fall into place. Thus, the general approaches on offer end up looking something like the following: demonstrate how moral rights for animals entail political rights for animals, or show how including animals in the social contract implies extending the political protections of that contract to animals, and so on. There are obvious limitations to such an approach, not the least of which are that it lacks a robust analysis of the links between ethics and politics and that it is effective only if a given audience shares the presuppositions of the specific moral theory under discussion.

I have been tempted to read Cavalieri's work up to this point as employing this same basic strategy of showing how adopting a particular moral theory has political implications for the treatment of animals, with the important exception that, in Cavalieri's case, she argues from the widely shared premises of a human rights framework rather than arguing that this framework is in fact the most coherent and logically defensible framework (as most of her analytic colleagues are wont to do). But her present chapter complicates this reading by making the case, along Critical Theoretical and Bourdieusian lines, that seeking rights for great apes and cetaceans has the potential to exploit contradictions in the dominant social and conceptual order concerning human-animal relations, thereby opening up further possibilities for more widespread destabilization of other hierarchical and chauvinistic ideas and practices concerning animals.

In short, then, Cavalieri's preferred strategy—if I understand it correctly—would be to pursue changes with regard to the standing of certain animals within the established socio-economic and legal order in view of those changes (once achieved), eventually producing a broader shift in the cultural imaginary toward animals more generally. Even though such struggles for rights only for certain animals might not appear particularly radical (in the sense of failing to institute a fundamentally different order for all animals), Cavalieri suggests that such reforms and initiatives can themselves be employed toward radical ends. Given the constitution of the current established order, and the ways in which various modes of anthropocentrism and capitalism have come to saturate social life at so many levels, I believe that Cavalieri's proposed strategy makes a great deal of sense. Her reservations about the likelihood of imminent and dramatic revolutions on behalf of animals are well stated; I would agree that those of us who struggle in view of creating a more just world for animals have little choice at present but to work within the interstices of the current established order, an order which is extraordinarily unjust toward animals but for which no radically different, widespread alternative appears to be on the imminent horizon. In such times, theorists, activists, and militants of all sorts have to be constantly vigilant for the kinds of possibilities that Cavalieri argues for, reforms that might appear minor at first glance but that could potentially unravel large portions of an unjust social fabric.[3]

My contribution to this discussion will not, then, take issue with Cavalieri's proposed strategy but will instead aim to complement it with additional perspectives and strategies directed at the kinds of radical ends we both desire. The supplement that I envision is touched on briefly by Cavalieri when she refers in her discussion of pro-animal feminism to the kind of work that needs to be done as a "precursor" to militant mobilization, work that she refers to as involving the "discursive identification of social actors" (Cavalieri, 31). Further in the text, I sketch out in more detail what this kind of strategy might look like when framed against the backdrop of an alternative analysis of what is at issue in struggles for animal justice. Before doing so, however, I should discuss briefly the position that Cavalieri attributes to me in her chapter for this volume, as there are important differences between the position that I wish to defend and the one with which she aligns me.

ANIMAL LIBERATION AND THE LEFT

In the theoretical schema of possible animal liberation strategies that Cavalieri develops, she aligns my work with activists who seek to address the limits of traditional, pro-animal liberalism (in the form of conscious consumerism and legal reformism) by way of a turn toward, and embrace of, the "leftist galaxy," by which she means neo-Marxists and new social movements that address injustices associated with race, class, and gender. The guiding idea here is not that pro-animal positions are already at work among leftists but that room can be made for animals within this leftist milieu. Cavalieri argues that this is not a particularly promising strategy, given the historical obstinacy of these movements to many of the theoretical and political advances associated with animal liberation. Now, it is understandable that Cavalieri attributes this kind of pro-leftist strategy to me, given a series of critical remarks I have made elsewhere about the need for larger numbers of mainstream animal activists and theorists to engage more directly with the specific problems posed for animals by the rapid spread of capitalism in its contemporary global, neoliberal, and biopolitical forms. Likewise, I have consistently aimed to demonstrate solidarity between animal justice struggles and many radical movements for social justice, especially those that take issue with the anthropocentrism of the established order. I should stress, though, that while I do not hold a strong *opposition* to those activists who seek to carve out space for animal issues within leftist struggles, it is by no means the chief strategy that I wish to advocate in advancing animal liberation.

Here, Cavalieri has attributed a position to me that is much closer to one advocated by Steven Best, who has argued for some time that animal liberation struggles should be seen as a continuation of the most progressive, leading-edge trends within the history of leftist struggles.[4] For Best, animal liberation should be seen alongside two other leading liberation movements, namely, human liberation and earth liberation. He argues that leftists (especially anti-capitalists within leftism) have made large strides in forming alliances with human liberation (in the form of social justice struggles and new social movements) and earth liberation movements (in the form of environmental justice and ecosocialist movements). But with regard to animal liberation, there has been little in the way of such linkages and alliances with broader leftist causes.

Best lays the blame for this failure on both leftists and mainstream animal advocates. In view of the latter group, he argues that mainstream animal advocacy has tended to reflect the perspective of white, privileged activists and organizations and, as such, has sought to remedy injustice toward animals primarily by way of demand-side, consumer-, and market-based solutions. While such advocacy is no doubt important insofar as it seeks to remove animals from the status of being commodities and property, it fails to take on board the broader leftist critique of capital and the systemic violence of contemporary capitalism's dominant modes and means of production.[5] Consequently, leftists have tended to see mainstream animal advocacy as a bourgeois pastime, a reformist project that is of interest primarily to those who have had their basic socio-economic needs met and who lack solidarity with those human beings who do not. Conversely, as Best explains, leftists have traditionally been slow to appreciate the moral strength of the argument for animal liberation (which is to say, the notion that the ideals of equality and liberation should be extended to include sentient animals) and its potential for being an important plank in the fundamental transformation of the established order. Although mainstream animal advocacy has indeed largely been captured by individuals and organizations who seek to work within the established socio-economic status quo, there has long existed alongside these groups a more radical collective of activists who understand well that animal liberation entails a fundamental challenge to the reign of capital in view of its incompatibility with the flourishing of much of both animal *and* human life. Leftists, Best maintains, should make common cause with this other, more radical approach to animal liberation that has been advanced primarily by underground, direct-action activists and their aboveground defenders.

Thus, Best aims to redraw the political strategic map by having us understand animal liberation as belonging properly within the orbit of the more radical wing of the movement, and then showing how that radical project for animal liberation has deep roots in the very values that the left has traditionally taken for its own. As Best writes:

> It is not understood by the Left or the animal rights/liberation move-
> ment, for example, that despite the amorphous political pluralism of animal
> advocacy, and absurd claims from some extremists on the Far Right to the
> contrary, the animal rights/liberation movement is fundamentally leftist
> in origins and values. The concerns for equality, rights, democracy, peace,
> justice, community, inclusiveness, nonviolence, and autonomy define both
> human and animal rights movements equally. The animal rights movement
> drank deep from the well of progressive modernism that also spawned radi-
> cal social movements, but hardly in a derivative and uncreative way that did
> not expand these values to their full meaning and potential.[6]

There is much to admire here in Best's strategy of linking animal liberation
with leftist movements along the lines of shared values and aims. It should
be clear that if struggles for animal justice are to gain additional ground
and force in the broader culture, they will need to form alliances with leftist
movements along these and other lines of affinity. Thus, as with Cavalieri's
proposed strategy, I find nothing to oppose in Best's approach. The question that should be raised here, though, concerns the prospects of making
such a strategy a central plank in movements for animal liberation. Cavalieri,
as I have already noted, sees little promise in trying to create alliances with
leftist movements that have proven repeatedly to be resistant to extending
basic ethical and political concerns to animals. Such alliances, she seems to
think, would amount to little more than animal liberationists begging for
support from groups that are largely hostile to pro-animal causes. Of course,
it might be said in response that this hostility from leftists derives not only
from a dogmatic anthropocentrism but also from a long-standing history
of mainstream animal advocacy that sees itself as a single-issue movement
divorced from other struggles for social and economic justice. Perhaps, if
animal advocates as a whole sought to broaden their focus and develop
genuine bonds of solidarity with other leftist movements, the prospects for
alliances along these lines would be somewhat better.

I would suggest, then, that Best's proposed strategy of building alliances
between leftists and animal liberationists is worth pursuing. But I would
also suggest that there are fundamental limitations to framing struggles for
animal liberation in this particular way, that is, as a deepening of what are
fundamentally Western, enlightenment values and a re-centering of radical
struggles on a neo-Marxist model. As is the case with many mainstream
animal philosophers and advocates, Best tends to explain violence toward
animals as an instance of speciesism and to see speciesism as the one of

the last, most stubborn prejudices that needs to be challenged by the ever-widening scope of movements for social and economic equality. Such an approach can, undoubtedly, help partially to illuminate the past and present status quo concerning animals, but, later in the text, I argue that it also masks a number of other dynamics at play in human-animal interactions as well as other strategies and frameworks for pursuing animal justice. My alternative approach begins with a critical analysis of the notion of speciesism. I then turn to a discussion of anthropocentrism and the kind of intersectional politics that I suggest might follow from a critique of anthropocentrism. In closing, I return to the question of capitalism in the context of this critical-political frame.

On the Limits of Speciesism

The critique of speciesism occupies a central position in much of contemporary animal ethics and animal studies. Despite widespread differences in the normative frameworks employed to argue for direct obligations for animals, nearly all pro-animal theorists agree that speciesism is the chief ethical and intellectual limitation that needs to be overcome in order to challenge the current status quo with regard to animals. Richard Ryder, who first used the term in 1970, characterizes speciesism as "discrimination or exploitation" against a being from another species that is defended "solely on the grounds" of that being belonging to another species.[7] As Ryder notes, "the speciesist regards the species difference itself as the all-important criterion."[8] Peter Singer, who is widely credited with popularizing the term, defines speciesism in line with Ryder as a "prejudice or attitude of bias in favor of the interests of members of one's own species and against those of members of other species."[9] The chief problem with speciesism, according to Ryder, is that "species alone is not a valid criterion for cruel discrimination. ... Species denotes some physical or other differences but in no way does it nullify the great similarity among all sentients."[10] In the briefest terms, then, speciesism is a prejudice that aims to justify discrimination and exploitation of members of nonhuman species simply because they are members of other species.[11]

The concept of speciesism is often fleshed out by comparing it with sexism and racism, which are considered parallel and paradigm forms of selfish and unjustified discounting of the interests of others.[12] Just as the racist or sexist discriminates against others on the basis of race or sexual difference, the speciesist discriminates on the basis of species differences. Such biological and physical differences (race, sex, species, etc.), it is argued by critics of these "-ism"s, are irrelevant to ethical consideration; as such, discrimination based on those grounds cannot be given cogent justification. Although various pro-animal authors define the term or expand on its meaning in slightly different ways, the basic definitions offered by Ryder and Singer as well as the parallels with racism and sexism recur throughout the primary texts of animal liberationists and animal rights theorists and activists.

I do not wish to undercut the important work that has been done around the concept of speciesism, for it has been enormously useful in helping to highlight some of the many ways in which animals have been discriminated against and subjected to harsh and unjustifiable violence. Along these lines, I am in fundamental agreement with pro-animal theorists who contest the dominant ethical order's attempts to place animals outside the scope of ethical consideration. I want to suggest here, however, that the concept of speciesism is perhaps not the most useful one for helping us gain a handle on the causes and nature of the marginalized status of animals in the dominant culture. Instead, I argue in what follows that the concept of anthropocentrism is a more useful term for defining the origins and nature of discrimination against animals, and that we would be wise to reorient animal ethics away from a focus on speciesism and toward a critique of anthropocentrism. First, we need to consider some of the critical limitations inherent to the concept of speciesism.

Common to much of the discourse surrounding speciesism is the idea that speciesism is best understood as an individual prejudice, attitude, or moral deficiency. To be sure, many individuals do, in fact, harbor negative attitudes toward animals that lead to their harmful or exploitative treatment. But, is the generally subjugated status of animals in the broader culture and the harm that is done to them best understood as the result of individual actions and moral deficiencies? David Nibert makes a compelling case that this understanding of speciesism betrays a confusion of ideology and system, or of prejudice and structure.[13] If we are seeking to understand the foundations on which animal violence is established and reproduced throughout the dominant culture, it is more helpful, he suggests, to see our individual attitudes and prejudices not as the primary cause of the problem, but as an ideological outgrowth of the institutional systems and economic structures that ground and frame our individual beliefs and actions. This kind of structural analysis does not deny individual agency, but it does shift the locus of where genuine power resides and of where efforts at transformation must be aimed. Individual attitudes and moral deficiencies undoubtedly need to be addressed, but they are not to be addressed simply through argumentation and education (which has, most often, been the dominant strategy of mainstream animal activists and organizations); instead, efforts aimed at fundamental changes in the basic structures and institutions through which we become individuals become the primary focus.

Nibert's point gains additional force when we think about speciesism as a parallel phenomenon with sexism and racism. In light of recent social science research, very few of us would argue that sexism and racism are best understood as resulting from individual moral deficiencies. To be sure, moral inconsistencies play some role in maintaining these stubborn problems, but we have learned to see sexism and racism more fundamentally as systems of power, with deep historical, linguistic, institutional, and economic roots. As such, individual changes in moral attitudes, while important, will not suffice to transform the larger structures and systems in which the institutional forms of racism and sexism reside. The same is no doubt true of the subjugated status of animals in our culture. The cultural and economic institutions through which violence toward animals is established and reproduced subtend and exceed our lives as individuals, and our personal moral prejudices tend to grow out of and reflect this system.

One of the effects of overemphasizing individual attitudes and prejudices in explaining the status of animals is that it leads us to believe that the chief means for addressing injustices are also to be found at the individual level. In the standard philosophical narrative, individual ethical prejudice is supposed to be overcome chiefly through an engagement with the philosophical arguments concerning the supposed "irrationality" of speciesism.[14] Unable to refute the arguments that demonstrate the inconsistencies of the speciesist attitude, as the standard narrative goes, an individual finds himself or herself moved by the force of reason to adopt an egalitarian animal ethic. From this nonspeciesist foundation, he or she is then further moved to make the kinds of changes in his or her personal life that remove speciesist bias: for example, adopting a vegan diet, buying cruelty-free products, refraining from visiting zoos, and so on.

The limitations of this individualist approach to addressing injustice toward animals become obvious if we recognize that the deep structures and institutions that create the conditions for the subjugated status of animals remain largely untouched by changes in individual consumption and behavior along these lines. Purchasing vegan food and cruelty-free products might send a less speciesist market signal, but such actions do not challenge the structural injustices of markets themselves, injustices that affect both animals and human beings in myriad ways. As I discuss in more detail below, the limitations of this kind of demand-side, market-based activism should encourage animal activists to forge links with a whole host of other justice struggles that are seeking to develop more just ways of living beyond the dominant capitalist vision.

If we continue this line of thought and turn our critical attention to an analysis of the dominant social and economic structures that give rise to the violent exploitation of animals, we also notice straightaway that their logic is not "speciesist" in any significant way. Those beings given full consideration by the dominant culture have almost never tracked with the lines demarcating biological species membership in *Homo sapiens*. There has always been unequal standing among beings who are considered full members of this biological group; even the most ethically humanist frameworks face intractable questions of what constitutes biological membership in the species *Homo sapiens* in view of beginning- and end-of-life questions, biological variations, hybrid species, and so on. In other words, the dominant culture and economy is not currently, and has never been, speciesist. As recent work on biopolitical approaches to animals has clearly demonstrated, what constitutes the human species has always been subject to divisions and ruptures of various sorts, and the placing of animals at varying points on this fragmented and divided terrain reflects less a kind of moral inconsistency and more the brutal lucidity of the machinations of sovereignty.[15] There are, of course, a handful of professional philosophers[16] who argue that ethical standing should track perfectly along the lines of biological species membership, but this is a rather contemporary view and does not reflect the historical and deep structural logic of the systems we are trying to understand and contest.

FROM SPECIESISM TO ANTHROPOCENTRISM

I would suggest that instead of speciesism, we think about the critique of violence toward animals and struggles for animal justice first and foremost through a critique of anthropocentrism. As I use the term, anthropocentrism refers to a set of ideas, structures, and practices aimed at establishing and reproducing the privileged status of those who are deemed to be fully and quintessentially human. Especially in Western cultural and intellectual traditions, the human (ὁ ἄνθρωπος) has only rarely and recently been taken to denote human biological species membership; more commonly, the human comprises only a select group of privileged individuals. Among those select individuals, humanness is maintained by seeking to exclude certain "nonhuman" traits and behaviors considered to be both internal and external to the human. What is included and excluded under the rubric of the human shifts over time, and group belonging expands and contracts, depending on a number of factors. One theme that remains deeply consistent, however, is the importance that the exclusion of animals and animality plays in relation to configuring the human. In my discussion of anthropocentrism here, I focus on those themes and features that help to highlight the relation between the human and animals/animality.

Anthropocentrism is premised chiefly on the notion that the human is *exceptional* among animals and all other beings. Characterized by purportedly unique capacities and powers, the human justifiably becomes the focus and center of attention. Thus, anthropocentrism is a kind of human narcissism, an attempt to grant importance, standing, and meaning to the human in nature and the cosmos. That there are points of continuity between human beings and animals has never been denied by the major intellectual and cultural sources of anthropocentrism. But what allows human exceptionalism and narcissism to be founded and maintained is the claim to a specific *anthropological difference*. Thus, even though humans share certain behaviors and characteristics with animals, anthropocentric logic posits that there must be a sharp ontological rupture where that which is specifically human emerges. Often, this rupture is attributed to capacities such as mind or language, or sometimes to a cluster of emergent traits, that give human beings their unique mode of existence.

This kind of exceptionalist ontology forms the deep background logic through which our normative, legal, and economic systems function. With the uniqueness of the human taken for granted, the circle of ethical consideration is drawn tightly in line with the contours of the human. Full consideration need not be given to all human beings within an anthropocentric context inasmuch as not all human beings are marked by the unique traits of "the human." Likewise, it would not fundamentally disrupt the logic of anthropocentrism to grant full or partial consideration to beings or species that are biologically nonhuman if this extension were based on their identity with, or interests of, "the human." Typically, however, full consideration is reserved for only a fraction of humanity, and even the most progressive humanisms do not usually argue for consideration to extend to *all* members and instantiations of *Homo sapiens* (despite rhetoric and protestations to the contrary). At an institutional level, the lack of any ethical standing leads to most animals and other nonhuman beings being reduced to property under the law and to commodities within the economy. Laws are designed with the interests of the human in mind, and markets are created and maintained to further the economic interests of the human alone. Although I am not concerned here with the Marxist question of whether the legal and normative practices of anthropocentric culture are simply ideological outgrowths of a capitalist economic base, it should be noted that anthropocentrism and the subjugated status of animals precede and exceed capitalist economies in many ways (historically, conceptually, ontologically, normatively, etc.). To be sure, capitalist economies intensify the effects of anthropocentrism at the institutional level and further entrench the deep background assumptions and ontology of anthropocentrism, but such economies can hardly be seen as the ultimate origin of the subjugation of animals and other nonhumans (we shall return to these issues in more detail in the closing section of the chapter).

So, what is to be gained if we take anthropocentrism as our critical point of departure? First, the concept of anthropocentrism better accounts for the subjugated status of animals as well as marginalized groups of human beings. The systems of power and control that structure modern societies are unevenly distributed among both human and animal populations, a fact that is easily overlooked if we think that speciesism is the main problem at hand. If we examine the world looking for speciesist limitations, we would expect to find individual ethical prejudices that operate according to a strict human/nonhuman species division. But, what we find instead are ideologies and institutions that grant full standing and privilege only to certain groups of human beings while excluding the vast majority of animals from consideration. In addition, anthropocentrism helps us better understand why certain animals are elevated over other animals, and even over other (often marginalized) human beings. Typically, the animals that receive privileged or protected status in our societies are those that are given that status for reasons germane to the interests of those deemed fully human (e.g., the elevated status given to "pets" that matter to people with standing). Such animals belong within the protected sphere of sovereign human interests and are seen as extensions and possessions of the human. Likewise, with those handful of animals for which full legal rights are sought (typically, mammals with higher-order cognitive abilities and subjectivity, such as the animals under discussion in Cavalieri's chapter for this volume), a certain form of anthropocentrism is often still at work in such instances. The animals that might be granted legal standing can only enter that sphere by way of strong analogy and identity with the human. Animals, along with human beings, who fall too far outside the scope of the dominant subject position of the human, are fated to various places, either on the edge or outside of the community of protected beings. It is precisely in view of such exclusions and marginalizations that we need to be particularly careful with regard to our strategies for seeking rights for animals, inasmuch as the criteria used to bring animals within the ethical, legal, political, and economic order will undoubtedly be found only among a small handful of animals; such criteria would likely also be used to justify excluding certain human beings and vast swaths of the more-than-human world from full protection.

At the normative level, the critique of anthropocentrism allows us to reframe the question of moral consideration. Critics of speciesism have often taken for granted that our current normative theories are sufficient for our interactions with other animals, as long as we are consistent with the basic premises of those theories and avoid excluding animals simply because they do not belong to the human species. Whether we adhere to consequentialist, deontological, virtue ethical, contractarian, or any of the other major normative theories, philosophers have been able to show that, in principle, these theories need not exclude other animals from consideration. As we noted earlier, that such frameworks tend to exclude some human beings and rather large numbers of animals from consideration is not necessarily a fatal objection for most critics of speciesism, as the aim of much of pro-animal ethics has been rigorously to avoid species-based discrimination rather than develop a broadly-inclusive framework.

The critique of anthropocentrism makes us skeptical of this approach to animal ethics. It would suggest that this kind of "new speciesism,"[17] which draws lines of consideration around select groups of human beings and animals, constitutes another iteration of anthropocentrism, with traditional human traits and capacities still occupying the center of ethical attention and consideration. A genuine challenge to anthropocentrism requires a rather different normative approach. Here, we would have to consider displacing "the human" from the central locus of the ethical project. Instead of a human-centered ethics that expands its circle of concern outward based on identity with the human, we would need to develop a broader framework in which humans are plain members in dialogue and relation with others of all sorts. It is beyond the scope of this chapter to lay out the details of such a framework, but the key point I wish to underscore is that the critique of anthropocentrism will be thoroughgoing only if its structural logic of inclusion and exclusion is challenged and replaced by a far more expansive mode of ethics.[18]

ANTHROPOCENTRISM AND INTERSECTIONAL POLITICS

If we take the critique of anthropocentrism (rather than the critique of speciesism) as one of our main points of departure, we can also start to reframe what is at stake in struggles for animal justice and how those struggles relate to, and intersect with, other movements for social justice. To begin, if we examine the situation of animals through a critique of anthropocentrism, we will gain a fuller sense of how the multiple lines of power and oppression that structure human culture also structure many forms of human-animal interactions; in other words, when we attend to such issues as race, class, gender, and colonialism in human relations, and the discriminatory practices aimed at people with different bodies and abilities or with nondominant sexual preferences or identities, we need also to appreciate the ways in which these modes of discrimination bleed into the dominant culture's relationship to animals. Animals suffer under (and, on rare occasions, benefit from) extensions and iterations of these same structures of power, and we, as animal advocates, must underscore that it is not some kind of stand-alone speciesism that explains the place of animals within the established order. Conversely, the analysis of how violence and power circulate throughout the animal world is an essential aspect of understanding what is often considered to be exclusively intrahuman violence. A wide variety of disciplinary apparatuses, technologies of control, and institutional logics of inclusion and exclusion have originated in human-animal relations; and it is rare that such modes of violence and sovereignty remain within the orbit of human-animal relations.[19] Anthropocentrism—as a logic and a rhythm of everyday life and as a system of power and practices—operates on and among bodies and lives of all sorts and does not respect any kind of speciesist logic.

What this analysis suggests, then, is that the various modes of oppression and structures of violence and power that have formed the critical focus of much of the activist and theoretical work of social justice struggles are themselves instances of anthropocentrism. Much as is the case with dominant forms of human-animal interactions, many forms of social injustice pass through and are caught up in anthropocentric logics that are predicated on human-animal and human-nonhuman distinctions; and the same holds true, conversely, in trying to understand many of the injustices suffered by beings in the animal and more-than-human world.[20] This is why it is essential to view anthropocentrism, rather than speciesism, as one of the primary sites of the critical focus of struggles for animal justice and social justice. I want to be clear that I am not arguing that radical social justice struggles are themselves simply instances of anthropocentrism; rather, I am suggesting that struggles against social injustice are actually struggles against anthropocentrism understood in the broad sense discussed above, and hence these struggles should be seen by those of us who work for animal justice as allied to our own movement. Although mainstream versions of social justice struggles often do tend to repeat uncritically anthropocentric logic and norms, the more radical and militant strains of social justice movements have long recognized that the marginalized and dispossessed are seen by the dominant order as not fully human; further, these more radical branches of social justice movements have emphasized that justice is achieved not by acceding to the privileged position of "the human," but rather by taking leave of this kind of hierarchical social order in favor of creating one that is truly universal. These radical, nonanthropocentric trends are found within nearly every major struggle for social justice, and they form fecund points of contact for building bridges between human, environmental, and animal politics.[21]

At issue, then, is the deepening of analyses and strategies of resistance that render manifest these interconnected and interlocking systems of power. What must eventually become axiomatic is the idea that if we are seeking to contest and transform power structures that encompass and control both human and nonhuman forms of life, then contestation of these structures that remains strictly at the intrahuman level is an insufficient strategy, inasmuch as it allows anthropocentric logics and practices to spread to other nonhuman domains, with the inevitable consequence of these forms of anthropocentrism rebounding back on marginalized human lives and institutions. Conversely, the notion that animal activists can focus solely on what is "good for animals" while ignoring anthropocentrism and social injustice in other registers has to be seen as being equally limited and problematic in perspective and strategy. As just noted, anthropocentrism operates on human and nonhuman territories in differential, interlocking, and self-reinforcing ways. Thus, to focus solely on speciesism and leave other modes of discrimination unaddressed (or, worse, to assume that they have already been essentially eliminated) is to commit a grave error that undermines struggles for both human and more-than-human justice.

In the past, some animal rights and liberation activists have glimpsed these connections to a certain extent. But, because the concept of speciesism was the guiding lens through which the connections were made with other forms of discrimination, the approach to making connections with other struggles was often based on an attempt to show that violence toward animals was the result of the same kinds of "irrational prejudice" that supposedly gave rise to sexism, racism, classism, and so on. Further, in order to make a compelling argument for animal liberation and rights, significant efforts were made to analogize and even strictly equate violence directed toward human beings and violence directed toward animals (the so-called dreaded comparisons).[22] Clearly, this strategy has not led to the desired results of creating lasting linkages between animal liberation and various human social justice movements and has instead tended to generate a backlash from the latter movements.

Animal activists and theorists might be tempted to see this backlash as a form of obstinate speciesism. I would suggest, however, that other dynamics are at play here. The attempt to show that violence toward animals is analogous or equivalent to racism, sexism, classism, and other forms of oppression can be carried out in a variety of ways. When (as has often been the case historically) these connections are drawn by activists and organizations within the animal liberation movement who are not seen as having demonstrated genuine solidarity with social justice struggles, the efforts tend to ring hollow. Also, if the equivalences and analogies are drawn in a quick, unthinking, or insensitive manner, they also tend to be rejected by those who struggle for social justice. When Cavalieri refers in her chapter to animal activists reaching out to social justice struggles and having to beg for support, we have to understand that this lack of support from other struggles often stems from a demonstrated lack of genuine solidarity for social justice on the part of the mainstream animal liberation movement more generally (I do not wish to paint with too broad of a brush here, and I am well aware that many activists do work in solidarity across various lines of struggle, a point to which I will return shortly). Large numbers of mainstream animal activists and organizations have often been entirely absent from the most pressing social justice and economic movements of our age. If, however, the animal liberation movement comes to be seen as fundamentally committed to various social justice struggles, both in word and deed, the responses from activists across the spectrum will undoubtedly be far more positive and supportive. Moreover, this kind of commitment on the part of animal liberationists will help to deepen and expand the understanding of how violence and power circulate among animal life and will create new critical and political perspectives. We cannot, of course, expect an outpouring of support for our causes or uniform solidarity in all instances. Anthropocentrism and violence toward animals contaminate the entire culture's thoughts and practices in countless and subtle ways that can be hard both to identify and to overcome. As such, even the most radical and critical struggles for social justice can reproduce certain anthropocentric logics and practices. But the possibilities for over-coming such limitations become far more likely if we are seen as being in genuine solidarity with those movements from which we seek support.

I want to emphasize that what I have just written about placing solidarity with social justice movements more in the foreground of animal justice struggles should not be taken to suggest that the issues of violence and control over animal lives are somehow secondary to dealing with human injustice. As pro-animal activists, we are certainly correct to insist that animals have suffered tremendously under the reign of anthropocentrism; even in those limited cases where animals have received privileged or protected standing, those positions have come at the cost of animals being subject to sovereign control in other problematic ways. Thus, to show strong solidarity with human social justice movements does not require us to give up our work or make it secondary in importance to any other cause; what *is* required, and what will make animal liberation and animal justice movements grow in force and numbers, is to inflect our work with a strong commitment to ending anthropocentric injustices in their many, interrelated forms. I should add that struggles for animal justice must, if anything, become *more* visible on the political scene, inasmuch as work that is done to break down hierarchical and violent human-animal divisions is exceedingly important for helping to highlight some of the more subtle and stubborn machinations of intrahuman anthropocentrism. By carefully examining the lives of beings who have been determined by definition and kind to belong permanently outside the orbit of the human, we gain a much clearer picture of the inner logic and problematic underside of discourses and practices aimed at protecting the privilege of the human. Thought and practice that proceed from within these critical spaces help give the lie to claims that the anthropocentric order can somehow be expanded or reformed to do justice to those who have traditionally been on the margins of the established human order. Anthropocentrism, as a logic and set of discourses and practices, continually produces violent zones of inclusion and exclusion—margins, thresholds, and outsides—that serve to shore up the sovereignty of the human. The point of genuine justice struggles, whether they are undertaken in view of marginalized human beings or beings that have been deemed sub- or nonhuman, is to contest the anthropocentric order of things and to create radically nonathropocentric forms of life.

This kind of broad nonanthropocentric project constitutes a continuation and deepening of historical and contemporary forms of theoretical and political activity. Perhaps, the strongest and most fruitful example of such work can be found, more broadly, among pro-animal feminists and ecofeminists.[23] What is particularly significant about the work carried out under this heading is that it is deeply intersectional in terms of its critique of power, simultaneously working along multiple lines of analysis. On the feminist side of this project, there exists a long history of feminist politics and theory that has emerged through serious engagements with a wide variety of other social justice movements. Currently, feminists widely accept the notion that the critique of sexism must, out of necessity and solidarity, involve the critique of a whole host of other forms of oppression; this is due, in large part, to the indefatigable work of women of color feminists who have made a powerful case for the irreducible necessity to think and act critically in intersectional terms.[24] This rich, intersectional vision of social justice has been subsequently expanded by ecofeminists and pro-animal feminists toward a consideration of the ways in which the logics and practices of social injustice are also at work in relations with animals and the nonhuman world more generally. In brief, using the terms I have developed above, ecofeminism and pro-animal feminism draw attention to the ways in which anthropocentrism creates caesuras and fractures internal to the human as well as in relation to what it takes to be nonhuman. These theorists and activists are urging us to understand these logics and their effects on both human and nonhuman populations as an interlocking, overlapping set of power relations that must be jointly critiqued and resisted.

Ecofeminism and pro-animal feminism have had mixed fortunes on the theoretical and political scene in recent years, with their presence going into decline for some time, but currently making a comeback in social justice movements, environmental justice movements, and recent work in animal studies and activism. Although I think we should strive to maintain the particular configuration and conjunction of feminist, animal, and ecological frames, I am not advocating we remain locked into this specific approach. Rather, I would suggest that the aim should be both to maintain ecofeminist/pro-animal feminist analysis while expanding that kind of intersectional approach toward other important social and environmental justice movements. We are currently seeing the formative stages of this kind of work in the expansion of critical animal studies (which builds on an already well-established anarchist and Critical Theoretical intersectional analysis and alliance politics) as well as in efforts to bridge pro-animal theory and activism with movements that have not seen much attention among animal liberationists, such as critical race theory, critical disability studies, and indigenous decolonial struggles, to name just a few.[25] These intersectional projects represent just a handful of the important efforts that are being made by theorists and activists to help the animal liberation and animal justice struggles mature into broad, powerful movements that offer an important critique of the dominant social order while also developing other forms of life beyond anthropocentrism.

CONCLUSION

As a means of bringing my remarks to a close, allow me to return to the theme with which I began, namely, the question of the left and its relation to animal justice struggles. In particular, I want to return specifically to the economic dimension of leftist struggles and their relation to animal liberation. As already noted, Cavalieri suggests that those of us who are interested in the conjunction of contesting economic injustice and contesting animal injustice have a fascination for the Left and believe that animal liberation can be achieved largely by expanding leftist struggles. There are, indeed, theorists and activists who defend positions similar to this one, but I hope to have made clear here that the left holds no particular fascination for me. Yet, given the widespread impact of leftist and progressive forces for socio-economic change, it would be foolish not to support these movements where appropriate and where they create possibilities for improving the conditions of life for human beings, animals, and the nonhuman world more generally. In brief, I am in deep solidarity with much of what leftist struggles are seeking to achieve, but I do not wish to characterize animal liberation in particular as the result of expanding basic leftist values to animals.

In my review of Peter Singer's edited volume *In Defense of Animals: The Second Wave* (from which Cavalieri is deducing my position), I did, indeed, stress the importance of animal liberationists paying more attention to capitalism.[26] I made those remarks in view of a volume which, while claiming to represent cutting-edge trends in animal defense politics, was nearly silent on the question of capital and instead sought to endorse a kind of demand-side, market-based approach to reducing animal cruelty (as an example, the volume features a rather uncritical interview with well-known libertarian and CEO of Whole Foods, John Mackey). This kind of economic conservatism, coupled with what I believe to be a naïve faith on Singer's part in democratic and legislative processes to improve the lives of animals,[27] led me to take a rather negative stance toward this vision of the future of animal liberation. Cavalieri is correct, though, to say that there is little use in fighting an abstraction such as "capitalism" when trying to address the economic problems we are dealing with in regard to animals. What I had in mind there is less a general critique of capitalism and more a set of strategies aimed at contesting specific aspects of economic exchange involving animals. Alongside vegan outreach efforts and other demand-side strategies (which are necessary but clearly insufficient), there is a need for large numbers of theorists and activists to be involved, for example, in contesting trade deals that negatively impact animals, workers, and the environment; similarly, efforts aimed at limiting subsidies to large animal agricultural corporations should be strengthened and supported; most important, it is essential that we support and help build alternative economies and forms of life that shrink and limit the influence of capital as much as is possible. Contemporary capitalism (in what we might call its global, neoliberal, biopolitical form) is deeply incompatible with the general flourishing of animal life and is predicated largely on the notion that animals and their habitats are nothing more than resources to be used in the interests of "the human." To believe that significant pro-animal reforms can be made to this system or to its legal and political apparatuses strikes me as fundamentally misguided. Instead, I believe our efforts would be better spent in supporting those cultures, economies, and practices that are aimed at a life beyond productivism and commodification, examples of which are proliferating among minoritarian struggles in our anti-capitalist, deanthropocentric, decolonial present.

How, then, might the ideas we have been developing here supplement and inflect Cavalieri's approach? Let us return to Cavalieri's strategy of securing rights for great apes and cetaceans. Beyond the necessity of instituting such rights within legal and political contexts, we would also have to consider the conditions necessary for those beings to flourish. In addition to threats from poaching and overharvesting, the primary threats facing great apes and cetaceans include habitat loss and degradation, climate change, and fundamental shifts in a wide array of other biophysical systems on which these animals depend. The analysis I have presented here suggests that many of these problems are driven, in part, by a kind of anthropocentrism, which treats the vast majority of the human and nonhuman worlds as resources for the benefit of a handful of persons. When this anthropocentric frame is combined with economic and political systems that intensify and radicalize this "world," the result is the kind of widespread and catastrophic environmental breakdown characteristic of our contemporary age. Fighting for rights for animals thus entails fighting against the present established (anthropocentric, capitalist) order and striving to institute other economies and modes of human-animal relations. At the same time, it is clear that this established order is at odds with the economic and ecological well-being of marginalized and dispossessed people and more-than-human beings of many kinds. Struggles for animal rights and animal justice can, thus, be situated alongside other radical social justice movements that address these issues as allies in the fight against a system of interlocking and mutually reinforcing oppressions. Animal justice struggles stand to benefit immensely from such alignments and alliances—and not only because our struggles are in need of more individuals who support them but also because such alliances and dialogues can help to illuminate the inner logic and structures of the anthropocentric-capitalist order. These kinds of intersectional, oppositional struggles are, however, not an endpoint; they must eventually give way to efforts aimed at rebuilding the world along different, nonanthropocentric lines. Such a project asks us to think differently not just about how the human is configured, but also about the status of human-animal relations and human-nonhuman relations more broadly. It is at this point that critical strategies open onto the field of political ontology proper, and where the radical displacement and decentering of the human might finally occur.[28]

NOTES

1. In response to Cavalieri's generous invitation to discuss animal liberation strategy, I should underscore my distance from standard ways of dividing strategy from tactics as well as from the traditional connections that have been drawn between avant-garde theorists and the development of strategy. When I refer to strategies in what follows, that concept should be read as operating on the level of what María Lugones calls "tactical strategies," that is, strategies that emerge from street-level resistance to oppression and that reject theorizing from above. The strategies that I explore and support here are ones that have emerged primarily among activists in a wide variety of radical justice struggles. Thus, what I have to offer is less a novel proposal for animal justice struggles and more of an endorsement and amplification of emerging trends among a variety of past and contemporary struggles. I make these

remarks at the outset to underscore the importance of thinking carefully about how we frame struggles for animal justice, the histories and movements with which we seek to build linkages, and the subject-positions of those who purport to speak and write on these issues. For more on the concept of tactical strategies, see María Lugones's essay, "Tactical Strategies of the Streetwalker/Estrategies Tácticas de la Callejera," in *Pilgrimages/Peregrinajes: Theorizing Coalition Against Multiple Oppressions* (Lanham, MD: Rowman & Littlefield, 2003), 207–237.

2. See, in particular, Paola Cavalieri, *The Animal Question: Why Nonhuman Animals Deserve Human Rights* (New York: Oxford University Press, 2001), 140; and, more generally, Paola Cavalieri, "The Death of the Animal: A Dialogue on Perfectionism," in her *The Death of the Animal: A Dialogue* (New York: Columbia University Press, 2009), 1–42.

3. This point should not be taken to imply that I think a realistic disposition leads to fatalism, nor should it be read as a suggestion on my part that radical tactics, such as direct action, should be ruled out. Following many activists, I believe a wide variety of tactics are helpful and necessary, and that tactics must be adapted to specific contexts. But, I do think it would be a mistake to assume that our struggles are part of an irreversible and inevitable teleological march toward animal liberation, as some mainstream activists suggest. When viewed in global terms, violence toward animals is at present increasing at exponential rates across multiple registers, and there is no guarantee that such trends will be reversed in the near future. As such, our proposed forms of resistance must take this rapidly changing and expanding reality into account.

4. Best has articulated his position over the years in a series of articles and talks, many of which have been helpfully collected in his recent book, *The Politics of Total Liberation: Revolution for the 21st Century* (New York: Palgrave Macmillan, 2014).

5. One of the most lucid articulations of this point can be found in Adrian Parr, *The Wrath of Capital: Neoliberalism and Climate Change Politics* (New York: Columbia University Press, 2013), chap. 6.

6. Best, *Politics of Total Liberation*, 102.

7. Richard Ryder, "Speciesism," in *Encyclopedia of Animal Rights and Animal Welfare*, ed. Marc Bekoff and Carron A. Meaney (Westport, CT: Greenwood Press, 1998), 320.

8. Ibid.

9. Peter Singer, *Animal Liberation* (New York: Ecco, 2002), 6.

10. Richard Ryder, *Animal Revolution: Changing Attitudes towards Speciesism* (Oxford: Basil Blackwell, 1989), 6.

11. Cavalieri uses the term *humanism* in her chapter to describe this phenomenon of speciesism. She notes that humanism is "the socially inculcated doctrine of human superiority warranting the discrimination against, and the

exploitation of, all the other animals" (Paola Cavalieri, "Animal Liberation: A Political Perspective," in this volume, 15).

12. Singer, *Animal Liberation*, 9.

13. David Nibert, *Animal Rights/Human Rights: Entanglements of Oppression and Liberation* (Lanham, MD: Rowman and Littlefield, 2002), chap. 1.

14. A version of this argument can be found in Peter Singer's "Foreword" to Cavalieri, *Death of the Animal*, ix–xii.

15. These points about biopolitical divisions internal to the human are developed most notably by Giorgio Agamben, *The Open: Man and Animal*, trans. Kevin Attell (Stanford: Stanford University Press, 2004); Roberto Esposito, *Third Person: Politics of Life and Philosophy of the Impersonal*, trans. Zakiya Hanafi (Cambridge, UK: Polity Press, 2012); and Cary Wolfe, *Before the Law: Humans and Other Animals in a Biopolitical Frame* (Chicago: University of Chicago Press, 2013).

16. In his analysis of speciesism, Singer notes that in order to avoid it, "whatever criteria we choose …, we will have to admit that they do not follow precisely the boundary of our own species" (*Animal Liberation*, 19). We should ask in response, though: How many people actually hold this kind of genuinely speciesist position that tracks precisely along the boundaries of biological species? Is this concept of an all-inclusive species membership notion of consideration actually the dominant logic of our main practices and institutions? It is true that at the rhetorical and conceptual levels, some philosophers and theorists have been speciesist (and, a careful reading of Singer's early work shows that it is contemporary speciesist philosophers who are his primary targets along these lines); but we are not simply trying to critique positions that philosophers hold in the abstract. We are trying to contest social and economic practices of power that have always given unequal and divided statuses both to animals and human beings and that do not track closely with concepts of biological species. The dynamics at work here are far more complex and are subject to continual reworking and transformation; to understand and contest them requires a critical framework capable of the same flexibility.

17. Joan Dunayer develops the concept of "new speciesism" in her essay, "The Rights of Sentient Beings: Moving Beyond Old and New Speciesism," in *The Politics of Species: Reshaping Our Relationships with Other Animals*, ed. Raymond Corbey and Annette Lanjouw (New York: Cambridge University Press, 2013), 27–39.

18. There are numerous versions of such an ethics found in the field of environmental ethics. I have offered an account of an ethics based on universal consideration in *Zoographies: The Question of the Animal from Heidegger to Derrida* (New York: Columbia University Press, 2008), Chap. 2.

19. See, for example, Joseph Pugliese, *State Violence and the Execution of Law: Biopolitical Caesurae of Torture, Black Sites, Drones* (New York: Routledge, 2013).

20. I am here underscoring a point made by Andrea Smith in her helpful essay, "Humanity Through Work," *Borderlands* 13 (2014): 1–17. Smith argues that settler colonialism is founded not just on a human-animal distinction but on a radical rupture between the human and the entire nonhuman world.

21. On this point, see David N. Pellow, *Total Liberation: The Power and Promise of Animal Rights and the Radical Earth Movement* (Minneapolis: University of Minnesota Press, 2014).

22. Marjorie Spiegel offered the original version of this approach in *The Dreaded Comparison: Human and Animal Slavery* (New York: Mirror Books, 1996). For an excellent discussion of the positive potentials and critical limitations of dreaded comparison analyses, see Kim Socha, "The 'Dreaded Comparisons' and Speciesism: Leveling the Hierarchy of Suffering," in *Confronting Animal Exploitation: Grassroots Essays on Liberation and Veganism*, ed. Kim Socha and Sarahjane Blum (Jefferson, NC: McFarland, 2013), 223–240.

23. The literature here is too extensive to cite. Instead, I point the reader to a recent and very helpful selection of essays from a variety of pro-animal and ecofeminist perspectives: Carol J. Adams and Lori Gruen, eds., *Ecofeminism: Feminist Intersections with Other Animals and the Earth* (New York: Bloomsbury Academic, 2014).

24. This literature is also too vast to cover here. In addition to Kimberlé Crenshaw's groundbreaking work on intersectionality, perhaps the most influential volume in this area is Cherrie Moraga and Gloria Anzaldúa, eds., *This Bridge Called My Back: A Collection of Writings by Radical Women of Color* (Watertown, MA: Persephone Press, 1981).

25. For a small sampling of this work, see A. Breeze Harper, ed., *Sistah Vegan: Black Female Vegans Speak on Food, Identity, Health, and Society* (Brooklyn: Lantern Books, 2010); Anthony J. Nocella II, Judy K.C. Bentley, and Janet M. Duncan, eds., *Earth, Animal, and Disability Liberation: The Rise of the Eco-ability Movement* (New York: Peter Lang, 2012); and Margaret Robinson, "Veganism and Mi'kmaq Legends," *Canadian Journal of Native Studies* 33 (2013): 189–196. The activist and theoretical work of pattrice jones has been especially influential along these lines.

26. Matthew Calarco, "Review of *In Defense of Animals: The Second Wave*," *Journal for Critical Animal Studies* 5 (2007): 82–87.

27. See, in particular, Singer's questionable criticisms of covert direct action in his "Introduction" to *In Defense of Animals*, which derive from his faith in achieving change through the democratic process: "In a democratic society, change should come about through education and persuasion, not by intimidation" (Peter Singer, "Introduction," in *In Defense of Animals: The Second Wave*, ed. Peter Singer [Malden, MA: Blackwell, 2006], 10).

28. Although I do not develop this point here, I want to underscore the notion that there are genuine limits to intersectional analysis and politics when it

comes to struggles for justice in view of animals and many other nonhuman beings. The notions of power (discipline, control, governmentality, etc.) that circulate in most forms of intersectional politics originate in anthropocentric contexts and do not always do full justice to the host of ways animals and nonhuman nature are captured by and escape the purview of various apparatuses. What is required here is a more detailed analysis of the ways in which sovereignty and animal/nonhuman life intertwine, of the sort offered by Dinesh Wadiwel in *The War Against Animals* (Amsterdam: Rodopi, 2015) and Mick Smith in *Against Ecological Sovereignty: Ethics, Biopolitics, and Saving the Natural World* (Minneapolis: University of Minnesota Press, 2011). I am also in fundamental agreement with Jasbir Puar's concerns about the tendency of intersectional analysis to lock subjectivities into their disciplinary modes (Jasbir K. Puar, "Queer Times, Queer Assemblages," *Social Text* 23 [2005]: 121–139). In terms of developing alternative forms of life beyond the capitalist-anthropocentric order, it is essential to think outside the constraints of the identities and subjectivities that we have inherited from the dominant tradition—hence the need to return to the field of political ontology. I have broached this topic at more length in *Thinking Through Animals: Identity, Difference, Indistinction* (Stanford: Stanford University Press, 2015), chap. 3. For more on the limits of intersectional analysis in view of the more-than-human world, see Nina Lykke, "Non-innocent Intersections of Feminism and Environmentalism," *Women, Gender, and Research* 18 (2009): 36–44.

Make It So: Envisioning a Zoopolitical Revolution

Sue Donaldson and Will Kymlicka

Paola Cavalieri's searching lead chapter in this volume asks "whither now" for the animal rights (AR) movement. Confronted with the ever-increasing scope and intensity of human violence toward animals, we must acknowledge the relatively insignificant inroads of AR activism to date. Cavalieri argues that this, in turn, requires acknowledging the inadequacies of standard liberal and left approaches to bringing about social change for animals, and developing alternative frameworks.

Our aim in this chapter is to provide a partial defense, and also a reformulation, of a specifically liberal approach to justice for nonhuman animals, and to identify its strategic resources for the AR movement.[1] Cavalieri is not alone in doubting the potential of a liberal framework for

For helpful comments on previous versions of this chapter, we would like to thank Paola Cavalieri, Darren Chang, Joost Leuven, Hilal Sezgin, and Carlos Tirado.

S. Donaldson • W. Kymlicka (✉)
Department of Philosophy, Queen's University,
Watson Hall 313, Kingston, ON K7L 3N6, Canada
e-mail: cliffehanger@sympatico.ca

P. Cavalieri (ed.), *Philosophy and the Politics of Animal Liberation*,
DOI 10.1057/978-1-137-52120-0_4

71

guiding a radical change for animals. Her chapter identifies and critiques two visions of liberal political strategizing. In the first vision, "a liberal stance instantiates conformity to the given," and "can merely produce requests for moderate measures that don't defy the status quo."[2] This stance assumes that existing institutions are based on "acceptable principles," which need only to be perfected, not overthrown (22). The second more radical liberal vision imagines a genuinely transformed polis. (She includes our own work in this second camp.) The problem with this radical liberal vision, however, is that it fails to "envisage the means." Lacking a suitable theory of "social agents, power relations, and strategies" (21), it amounts to a strategy of passively awaiting history, or crisis, to usher in a new world.

These limitations of liberal democratic politics help to explain, in Cavalieri's view, the slow progress of the AR movement. On the one hand, it has pursued ad hoc, deficient (welfarist), regulatory reform—a strategy that is directed at legal and institutional structures, but espouses deeply inadequate and nontransformative goals. On the other hand, considerable AR liberal activism has focused on an individualist, vegan, consciousness-raising strategy ("a gradual illumination of minds"), and a related "conscious consumer" strategy, which may espouse radical goals for animal liberation, but which lacks any direct challenge to institutions and structures.[3] The first is deficient in aim, the latter in strategy.

The radical left, which pursues anti-system animal activism such as calls for a grand alliance against capitalism, suffers from its own deficiencies. Cavalieri argues that it fails to account for the deep indifference to the animal question on the left, and the danger and/or futility of advocating revolution under current conditions.

In place of these failed liberal and left approaches, Cavalieri proposes strategies founded on a combination of Frankfurt School critical theory and Bourdieusian reflexive sociology, both of which aim to disrupt the ways in which oppression becomes naturalized as part of our taken-for-granted "habitus." As she puts it, "political subversion presupposes cognitive subversion, and in particular the unmasking of the process of euphemization through which the dominant make violence unrecognizable by conferring on it the appearances of the natural" (31). The role of theory, she suggests, is not to "stick to the regulative ideas offered by the soundest declension of universalizability and the best liberal utopias" (28), but rather to enable the sort of cognitive subversion of the habitus that can lead to political subversion. She then discusses how the strategy of pursuing legal rights for great apes and whales can be a politically effective form of cognitive subversion, undermining the human supremacist ide-ologies that naturalize violence against animals, and prefiguring new forms of human-animal relations.

We share Cavalieri's concern about the limits of existing AR strategies, and about the importance of subverting the taken-for-granted habitus. Indeed, our own work has been inspired and shaped by her work critiquing human supremacist ideologies[4] and by her arguments defending the rights to personhood of great apes and whales.[5] So, we write from a position of overall agreement. But, we would offer two amendments to her theory: the first concerns liberal-democratic politics; the second concerns subvert-ing the habitus. These amendments are more a matter of emphasis and nuance than deep disagreement on principle, but they may, nonetheless, be significant for how we envision AR strategy.

RETHINKING THE HABITUS

Our first argument—and the main focus of this chapter—concerns the potential of liberal democratic politics, and in particular its potential for social justice movements. Drawing on historic examples, we will offer a description of how radical social transformation can be achieved in liberal democracies, and why, in our view, this possibility extends to the animal question. If the AR movement has failed to date, we will suggest, it is not due to the structural limitations of liberal democratic politics or liberal political theory, but rather because the AR movement has not taken full advantage of the possibilities such politics offer. In any event, it is prema-ture to declare the failure of radical liberal politics when its possibilities have not yet been adequately tested and explored.

Before turning to that argument, however, it may be useful to flag our second point of disagreement, concerning how we conceive the task of subverting the habitus, and of denaturalizing human violence against animals. In Cavalieri's view, this process is, in the first instance, one of conscious reflection. As quoted earlier, she starts from the assumption that "political subversion presupposes cognitive subversion" (31), and so the task is to challenge the conscious ideologies, such as human supremacism, that sustain animal oppression.

Recent findings in social psychology suggest, however, that the "alter-ing of cultural hegemony" (30) is not only, or even primarily, a process of changing people's conscious beliefs. The habitus is equally a matter of embodied habits and unreflective knowledge, belief, and behaviors. Much of our behavior, including our behavior toward animals, is prompted directly by emotion or by intuition, unguided by conscious reflection, and/or consists of instilled habits and adherence to social norms that are triggered by social cues that we are rarely conscious of, and even more rarely subject to conscious reflection.[6] This suggests that one effective way to alter cultural hegemony is to directly disrupt people's embodied habits. Consider the following example. Let us imagine that Stan has grown up with (and been socialized into) "common sense" assumptions that animals exist to serve human ends, that humans need to eat meat to be healthy, and that animal use is necessary and unproblematic. This common sense understanding has been constantly reinforced by the casual indifference to animals' welfare built into countless dimensions of human society, the ubiquitous presence of flesh foods and other animal products and their uses in everyday life, and the presence of innumerable humans who engage in the same practices as Stan without any apparent thought or crisis of conscience. Now, one way to disrupt this habitus is intellectual disruption, demonstrating to Stan that the human domination/exploitation of animals lacks a sound intellectual justification—it is neither necessary nor morally justified, but rather normalized and naturalized violence. This is the strategy of ideological critique, which underpins the Frankfurt School's vision of critical theory.

A different approach to changing Stan is to change his environment. Let's say someone at the prison where Stan works starts putting up special film on the office windows to deter bird strikes. Then, vegan food options start appearing in the prison cafeteria, as well as at nearby restaurants and grocery stores. Meanwhile, the local town council decides to ban the use of wild animals in circuses and pet sales. Local planners start making design decisions to better accommodate urban wildlife, and to promote biodiversity. Some of Stan's acquaintances become more assertive about bringing their animal companions to work or on family holidays, or allowing them on public transportation, or establishing dog parks. Perhaps, a small farmed animal sanctuary is established at Stan's workplace as part of the rehabilitation process for offenders, who help care for the animals (e.g. by looking after stray dogs and cats who are put up for adoption, or by caring for sheep and goats whose day job is to help to control vegetation in the local park).

None of these changes is in itself revolutionary, and each is the result of multiple motivations (human convenience, physical and mental health, cost savings), some of which have little directly to do with AR. Yet, each in its own way disrupts habits and practices by which violence against animals, or indifference to their well-being, is normalized and rendered invisible in our society. And so, over time, Stan's common sense habitus is gradually, but profoundly, altered. His idea of who animals are, what they need, and the possibilities for interspecies social relations are transformed, not as a result of being exposed intellectually to ideological critique, but through lived experience. He may not even be consciously aware of these changes, perhaps, until a friend or family member says something like "Gee, Stan, I see you've shifted from using a lethal mouse trap to a humane trap. When did you become such a soft touch?" Or "Gee, Stan, you used to be completely cynical about young offenders, and now you can't stop posting photos of them with those crazy goats." Or "Gee, Stan, you used to be scathing about vegetarians but didn't I see you at that new vegan burger joint?" At these moments, Stan may become aware of how much he has changed, and wonder how it happened. It was not because Stan stopped to consider his views, subjecting them to ideological critique. Rather, change has come because Stan's environment has changed, and new ways of being have become real, viable, attractive, and self-reinforcing for him.

Undoubtedly, the AR agenda will require ideological critique as well. We discuss that part of the story in the following section. But citizens may be more open to the ideological critique—and to compelling new ideas and images of interspecies relations—when their lived environment has already changed, or is changing, to disrupt instilled practices and habits and to render more salient alternative ways of living. Moreover, there is good reason to believe that cognitive subversion is not sustainable or efficacious in the absence of broader changes to the habitus. Consider a recent study by the Humane Research Council, which finds that in the USA, only one in five vegans/vegetarians sticks with the diet, and most backslide within three months (2 % of Americans are vegetarians; 10 % are former vegetarians).[7] This suggests that the AR movement has already had some success in generating cognitive dissonance and subversion. After all, 12 % of Americans made a conscious commitment to reject the assumption that eating meat is justifiable or desirable, thereby proving their ability to resist the mountains of pro-meat propaganda from industry, media, and government. But, in most cases, this cognitive subversion is not sustained, primarily for nonideological reasons, such as a feeling of social awkwardness, the desire to avoid conflicts with friends and family, and the difficulty of finding vegan food in grocery stores or restaurants.[8] People do not revert to meat eating for ideological reasons—they do not suddenly reconvert to ideologies of human supremacism—they simply find it too difficult to follow through with their ethical commitments in a context where none of the everyday unreflective habits and social cues of their lived environment supports those commitments.

In other words, the AR movement has had some success in critiquing the ideological dimensions of our habitus, but has failed to create an alternative habitus in which AR commitments are supported and normalized at the level of everyday unreflective behaviors and embodied practices. The evidence suggests that change through cognitive disruption may be unsustainable without broader changes to people's everyday lived environments. An advocacy model centered on changing the conscious cognitive beliefs of individuals may therefore be a Sisyphean task, and one that sets up many (not all) individuals for failure and frustration.[9] Most people need supportive environments, communities, and institutions to be able to develop and maintain an animal-friendly way of life. It's possible that (currently) successful vegans belong to a small minority of the population whose behavior is more susceptible to conscious control, or who derive psychological benefit from being social outliers, or whose personal identities are less tied up in meat-based traditions. So, we cannot extrapolate from what works for the consistent 2 % of the population who are vegan. We need to consider whether very different approaches are necessary for the 98 %.

The process of changing social norms over extended periods involves all dimensions of our inhabited environment, not just ideas or mediatized culture. For academic theorists, it is an occupational hazard to think that most people change by reading and thinking, forgetting that for most of us, doing, being, and experiencing in less intellectualized ways may be more important. Strategic action and organization for the AR movement need to capitalize on this full spectrum of human ways of being. Instead of committing all of our energies into showing people that their beliefs are inconsistent or rationally indefensible, we must also work on changing people's experiences and environments, and making alternative ways of living a possible, normal, commonplace, and, eventually, compelling option. This, indeed, is the focus of many grassroots activists, but the evidence suggests we need much more of it, and that it needs to be given a more central role in our theories of AR advocacy.

One of our amendments to Cavalieri's picture, therefore, concerns the strategic importance of attending to the noncognitive and unreflective dimensions of our habitus. This also has implications for our primary argument related to the potential of liberalism. As noted earlier, Cavalieri argues that liberal approaches fall into two camps: first, those that uncritically accept existing institutions as legitimate; second, those that are naively utopian, disconnected from any plausible account of political agency and political change. We agree that it is not enough simply to articulate "the regulative ideas offered by the soundest declension of universalizability and the best liberal utopias," and that we need to think about how AR theories of justice and rights can empower political practice. But, we would emphasize that sustainable political mobilization for progressive change, like sustainable individual ethical commitment, must be rooted in supportive social environments. Dauvergne and Lebaron refer to this as the "infrastructure of dissent," and argue that it must be rooted in the "social texture" of people's lives—the dense field of shared habits, experiences, images, sensibilities, solidarities, dispositions, virtues, affiliations, loyalties, identities, and narratives that enable progressive groups and organizations to advocate for unpopular views and to resist pressure to conform.[10] So, we need to evaluate our AR theories, at least in part, by asking (1) whether they resonate with, and take advantage of, the social texture of existing infrastructures of dissent; and (2) whether they would embed animal concerns into this social texture, so that these concerns became a habitual part of people's everyday lived experience.

We believe that it is a distinctive virtue of our liberal political theory of AR that it fulfills these twin goals. It draws on the deeply felt images and narratives that drive contemporary progressive politics, and would, over time, incorporate concern for animals into the everyday lived experience of progressive movements. We offered this theory not just as a philosophical ideal or aspirational endpoint—not just as "the soundest declension of universalizability and the best liberal utopias"—but as a way of connecting diverse forms of animal activism to the familiar motifs and touchstones of contemporary progressive politics, and to the social texture and "social imaginary" of existing infrastructures of dissent. Lawyers, educators, legislators, AR campaigners, and others need to tap into this collective imaginary to be effective allies for animals, and to gain traction against the powerful interests that defend animal exploitation. Insofar as *Zoopolis* links the fight for animals to a larger social imaginary of progressive politics, and helps to put concern for animals at the center of the lived experience of these movements, then it may not be as deficient in strategy as Cavalieri suggests.

Or so we will argue in the rest of this chapter. We will also suggest that Cavalieri's own preferred approach may fall short in these strategic respects. While we fully endorse the philosophical argument for recognition of the rights of personhood of great apes and whales,[11] we're less sure that prioritizing these particular claims within the AR movement will engage our social imaginary in the right way, or change our everyday lived experiences and environments in ways that are most needed for progress.

Cavalieri thinks that recognition of claims to personhood of great apes and whales would stand as a particularly powerful antithesis to ideologies of human supremacism. And perhaps it would. But it may not stand as a particularly powerful way of engaging the narratives and practices of contemporary progressive politics or of incorporating animal concerns into them.

THE RADICAL POTENTIAL OF LEFT LIBERAL POLITICS

Cavalieri contrasts revolution, "a radical break in the course of history," with the tepid politics of liberalism and its "awkward path" to radical change. She notes that the working classes needed the French Revolution "to cease to be considered merely as a means for supporting some ruling classes" (21). Many AR theorists have made related claims about the inadequacy of liberal democratic politics for liberating animals, either because liberalism is deemed inseparable from capitalism, which is a root cause of animal exploitation[12]; and/or because a liberal democratic state is deemed inherently unjust or imperialistic. We have attempted to address some of these critiques of liberalism elsewhere.[13] In this chapter, our approach is more constructive, elaborating our positive story about how we see the potential for radical progress unfolding within the framework of liberal democratic politics.[14]

In thinking about the potential of liberalism and liberal democratic politics, we need to avoid twin determinisms. On the one hand, some people take contemporary neoliberal economics, with its market fundamentalism and its obsession with growth and capital accumulation, as somehow the inevitable consequence of liberalism. But this misconstrues the liberal tradition. Various strands of liberalism, including the left liberal strand that we endorse, have offered an extensive critique of capitalism. John Rawls, who has had such a profound influence on liberal egalitarian thought, argued that liberal justice is incompatible with laissez-faire/neoliberal capitalist institutions, or even with welfare state capitalism. He believed that some form of liberal socialism or "property-owning democracy" was necessary for the creation of just institutions. This is a far cry from any sort of acceptance of the underlying principles of existing institutions. In recent years, there has been extensive debate among theorists of the liberal left and the socialist left concerning the role of markets, and the nature of property ownership, in a just society.[15] We will not wade into that debate here. Suffice to say that for all their remaining differences, there is significant agreement among these theorists that just institutions entail the radical curtailment of economic inequalities, of concentrations of economic power, of the scope and freedom of markets, and of the influence of money in politics.

As Simone Chambers has noted, Rawls' property-owning democracy may have seemed a reasonable aspiration for Western democracies in the 1970s, but, in the current context, it looks politically radical.[16] The last 35 years have witnessed the ascendency of neoliberalization, and an erosion of the broad postwar welfare state consensus. But, it would be incorrect, in our view, to look at this development as an inevitable trajectory of the liberal state or liberal political economy. Western democracies did not need to destroy the advances made in the postwar period toward greater equality, and they did not need to let democracy slide into plutocracy (governance for the 1 %). Nor were they compelled to structure the economy on endless and unfettered growth, rather than on a rational awareness of finite resources and a fragile planet. Nor was the path to intensive animal exploitation an inevitable unfolding of a logic of capitalism or enlightenment rationalism. These developments were the (frustrating, infuriating) outcome of political contestations, not the "logic" of liberalism.

At the same time we must also avoid the opposite determinism, which assumes the inevitability of an ever-expanding "rights revolution." As Bonnie Honig notes, some liberals are prone to simply wait for the unfolding of a "chrono-logic" of the liberal "rights machine," as though the reasonableness of ideas of egalitarian justice has political agency in the absence of a political struggle for power. Ideas of a logical unfolding of rights too often displace the realities of political struggle:

> Looking backward, we can say with satisfaction that the chrono-logic of rights required and therefore delivered the eventual inclusion of women, Africans, and native peoples into the schedule of formal rights. But what actually did the work? The impulsion of rights, their chrono-logic, or the political actors who won the battles they were variously motivated to fight and whose contingent victories were later credited not to the actors but to the independent trajectory of rights as such?[17]

So, neither neoliberal market fundamentalism nor egalitarian rights revolutions are predestined, as if one or other were written into the DNA of all liberal societies, inevitably fated to emerge. We need to set aside such trans-historical and deterministic metanarratives and look at "social agents, power relations, and strategies."

Progressives have lost many battles in recent years, but not all of them. So, first, we need a careful historical analysis of successful social justice movements and how they might be relevant for animals. This requires that we overcome historical amnesia, and avoid what ecologists call shifting baseline syndrome (the failure to recognize that the norms one has grown up with represent a radical change from the experience of earlier generations).[18] Western societies have previously been through cycles of extreme income inequality, unfettered capitalism, and predatory global trade. Workers successfully organized for meaningful forms of regulation, and achieved a significant redistribution of power, especially in the social democratic states of northern Europe. These gains have been eroded in recent decades, and the battle for economic justice has been set back. But this is an outcome of struggles for power among individual and institutional actors, not an inevitable denouement of market-economy liberal democracy. Moreover, people are finding ways to fight back for their economic and social rights against neoliberal policies and institutions.[19]

In many areas, socially egalitarian and inclusive norms have achieved widespread acceptance despite the ravages of neoliberalism. The lives of countless women, children, GBLTs, racial and ethnic minorities, people with disabilities, and indigenous peoples have been dramatically transformed in recent decades. These struggles are ongoing, but it would be irresponsible to discount the advances that have been achieved (not to mention deeply insulting to the generations who have organized, fought, and won crucial battles for social justice). We need to recognize and protect the advances that we make, not trivialize (and risk losing) them, in our frustration that justice remains so painfully incomplete.

The social justice advances of recent decades have been achieved by processes of gradual reform in liberal democracies. In most cases, there is no crystalizing moment of revolutionary violence or radical rupture, and, perhaps, this absence of defining moments contributes to the historical amnesia that diminishes the significance of these struggles. There is an extended arc of moments—protests, boycotts, occupations, divestment campaigns, civil disobedience, legal cases, suppression and reaction, endorsements, creation of new public narratives, icons, and norms—which gets lost in the rear-view mirror in ways that the upheaval of a French revolution did not. But make no mistake, these are battles for genuine social transformation. In Joe Hoover's terms, these rights struggles are contesting the basis of legitimate authority and the boundaries of community: they are a demand not only for inclusion in existing social structures, but also for inclusion as a participant empowered to remake those structures.[20]

Historical amnesia leads us to underestimate the contestability and contingency of taken-for-granted rights. It is a mistake to focus solely on how liberal rights function once they have been legally instantiated (leading to misunderstandings of rights as largely axiomatic, hollow, or as tools of state power). To be sure, rights can become tools of governance, but they have also served as tools of resistance for the excluded to challenge existing structures of power.[21] Rights are claims—particularly powerful ways of framing moral and political claims as part of the struggle for inclusion and social change. It is their role as political tools for those excluded from the demos that is relevant for the AR movement.

This may sound rather abstract, so let us consider a specific example; Jerry Anderson argues that child labor reform in Britain during the transformations of the Industrial Revolution is a particularly illuminating comparison for understanding the potential to advance new moral claims for animals.[22] Before the Industrial Revolution, children worked on farms and in family businesses, but industrialization brought a massive deterioration in their working conditions. A "race to the bottom" in a ruthlessly competitive environment led to "longer hours at lower wages, increasingly poor accommodations and conditions, and increasing reliance on the cheapest laborers, i.e. children."[23] Similar to animals in the recent growth of industrialized agriculture, nineteenth-century child laborers were a vulnerable group caught up in new industrial processes under conditions of unfettered capitalism, limited in their ability to resist or organize against their circumstances, powerless in the political process, and short of allies (including their parents, who depended on their children's wages to ensure a subsistence family income). As the horrific magnitude of their conditions gained visibility, industrialists fought any efforts made by reformers, using the same free-market rhetoric that animal exploiters currently use regarding economic necessity, foreign competition, and capital mobility, and relying on the same lies concerning welfare provisions and humane conditions for those caught up in the system.

The battle to reform child labor conditions took more than 75 years within the UK and other liberal democracies, and continues internationally.[24] It was an incremental process with gains and setbacks as a new set of norms and meaningful legal protections were gradually established. Anderson considers it a prime example of a moral reform movement, one in which protection is gained for a powerless group "even when reform runs counter to the economic interests of those in power."[25] It is particularly relevant for

AR since children themselves were unable to self-organize, and so relied primarily on the advocacy of others. It, therefore, offers an important exception to Norm Phelps' claim that AR is the "first social justice movement in history" to operate without the organized, conscious participation of the victims of injustice.[26]

Anderson's outline of the key stages in the development of a moral reform movement can be summarized as follows:

1. A period of "competitive deterioration" leads to drastically increased exploitation. This generates demands for reform, which is initially forestalled by industrial interests in the name of wealth generation and unfettered markets.
2. Pressure for reform grows, increased by trigger events (e.g. disasters), popular culture (e.g. novels by Dickens), and moral leadership by key figures.
3. Interest groups form and start putting collective pressure on legislative bodies. Progressive theorists articulate ideas to counter market ideology. Consumer actions (e.g. boycotts) help to change the economic equation and broaden the constituency for reform. Together these activities contribute to the development of a new norm and legislative reforms.
4. Initial reforms are modest and fragile, subject to intense industry blowback, loopholes, and claw backs. Reformers must maintain pressure over many decades to solidify and build on initial modest protections. As industry loses ground domestically, it will attempt to regroup by exporting exploitation and developing alternative markets.[27]
5. Even when effective domestic (or international) reform is achieved, it must be constantly monitored, "like a cancer in remission," as long as economic pressures for deregulation exist.

As Anderson notes, the process by which a "new ethical/moral imperative develops" is a "complex subject."[28] In relation to animals at present, for example, stories about farmers who "care about the animals" are reinforced by a legal framework, which exempts them from anti-cruelty laws.[29] It takes multiple tellings of new stories to replace these entrenched understandings. And new stories are more powerful if they can connect to existing frames—for example, by linking animal protection to ideas of democratic renewal, protection of the vulnerable, and social reciprocity.

Moral entrepreneurs (leaders of stature and seriousness) can dramatically influence the pace with which new norms are adopted.[30]

These norm leaders can include creators of popular culture. But given the length of time required to bring about a shift in norms, organizational structure for a reform movement is crucial—a structure which allows for sustained pressure, and for continuation of strategy and aims despite changes of individual involvement. Norm change and incremental regulatory change create an iterative loop (as long as pressure for reform is maintained):

> The child labor history teaches that legislative reform will probably be a gradual process of establishing footholds, which slowly help to re-adjust the accepted norms. Thus, there is a feedback loop between government regulation and norm development.[31]

Similarly, consumer action tends to work in an iterative process with legal change, rather than being a significant driver of reform on its own. There is a collective action problem in persuading individuals to make consumer choices (e.g. boycotting animal products) if they perceive that they are making a sacrifice that others are not required to make. Evidence suggests most individuals may be more willing to support legislation that requires uniform change, rather than taking individual action.[32]

The end result of this complex process was the enshrinement of a set of children's rights that was revolutionary compared to the original baseline, but which most societies now accept as obvious and unquestioned. We should not generalize from a single example such as the child labor movement of the nineteenth century, and we do not have the space to review other relevant comparisons, but we would argue that the history of social justice movements suggests a number of important lessons:

1. liberal democratic rights have been a vital political tool in struggles of marginalized and dominated groups (even the most powerless) to win protection and power to transform the demos;
2. such incremental but transformative change is often overlooked due to historical amnesia and shifting shifting baseline syndrome;
3. successful incremental reform requires an organized social movement, which is able to institutionalize and sustain pressure on the status quo over an extended (indeed indefinite) time period, first to change norms, and then to support new norms; and

4. the process of gradual norm change is complex, involving iterative feedback loops between legal change/incentives, and public sensibilities/public pressure.

Finally we have complicated the final point by noting that, in light of recent developments in moral psychology, we should keep in mind the following lesson:

5. venues for influencing public sensibilities include indirect focus on people's lived environments and implicit learning and habit formation, as much as direct targeting of their conscious knowledge and belief structures.

In the remainder of the chapter, we will consider the implications of these points for how to think about strategy for the AR movement.

Expanding the Movement

To make progress toward justice for animals, AR advocates need to create a broad-based, sustained social movement. Transformative change requires institutionalization (not just individual conversion) across a range of social, political, and economic locations. This is what provides the staying power to maintain pressure over the long-term required to change social norms, and to revise state and regulatory structures. To do this, we need to be much bigger—drawing individuals into the movement, developing strong organizational structures within the movement, and making connections outside the movement with other progressive organizations.

We believe that AR theory and advocacy, in its current form, throws up many unnecessary barriers to the expansion of the movement. We will consider two broad concerns: first, how standard AR arguments and campaigns fail to reach out to other social justice movements, and may indeed repel them; and second, how standard AR advocacy often fails to reach out even to other animal advocates. In each case, we will suggest that a left liberal Zoopolitical conception of AR can help overcome these strategic barriers.

Linking AR to the Broader Family of Social Justice Struggles

To be successful, the AR movement needs to forge alliances with other progressive causes—activists working on disability rights, women's rights,

LGBT rights, anti-racism, multiculturalism, children's rights, environmental issues, grassroots democracy, food security, public health advocacy, and economic justice. We do not need to share perspectives all the way to the ground with other groups in order to work together. But we do need to nurture a sense of belonging to a larger community of progressives: we need to think and act as a member in good standing of the family of social justice struggles, and to be recognized as such. As Norm Phelps notes: "animal rights is consistent with the fundamental philosophy of the left; it is inconsistent with the fundamental philosophy of the right .... I think it is very important that we consider ourselves a social justice movement among other social justice movements, and that we work very hard to form a unified front with the political left."[33]

Unfortunately, at present, AR advocates are too often the "orphans of the left,"[34] alienated from, and disavowed by, other social justice movements. There is much to be said about how this orphaning came to pass. Attempts by governments and industry to smear animal activism with the label of terrorists and extremists are one factor, breaking potential "ties of generosity and solidarity" as other social justice movements seek to avoid any affiliation with AR groups that might "sully their reputation, frighten their donors, or endanger their ongoing (although much reduced) campaigns and membership."[35] Another part of the story, as Cavalieri notes, is that many people on the left are themselves deeply invested in ideologies of human supremacy and in the (often very personal) pleasures that arise from exploiting animals.

So the larger social and political environment is not particularly hospitable to AR advocates, and good-faith efforts to reach out to other social justice movements may well be rebuffed. But it is also true that AR advocates have not reached out as much as they should. Indeed, AR theorists and activists have often framed arguments and campaigns in ways that are at best strange and uncompelling, and at worst downright disturbing to progressives working on other social justice issues.[36] If we want to be accepted as members in good standing of the family of social justice movements, we need to think carefully about how our arguments and actions impact other social justice movements, and how we can strengthen ties with them, rather than acting as go-it-alone orphans. This requires changes at two different levels: first, changes in the strategies we use to pursue AR goals; second, and perhaps more interestingly, changes in the very way we define AR goals.

Consider first the strategies we use to advance AR goals. To take an obvious case, if we want to be accepted alongside feminist groups as members of a common family of social justice movements, then we clearly need to avoid sexism in our tactics (as well as in AR organizational hierarchies). Yet as many commentators have noted, some AR groups have engaged in such tactics, the sexist campaigns of PETA being one example.[37]

Another problem is the manner in which some AR organizations have decided to play the corporate game. This is evident not just in choices about media tactics (shock tactics, celebrity campaigns, sex sells) but also in organization structure and governance, such as putting emphasis on organization's fundraising and growth, and on fitting into the neoliberal economy, rather than trying to use the power and resources of the organization to anchor alternative economies and practices. If AR organizations act like corporations—being secretive about their practices and policies, exploiting interns and low wage employees, and engaging in capital flight rather than anchoring their work in local communities where they can provide stable employment—then we cannot be surprised if workers and social progressives do not perceive AR as a social justice movement.[38]

Similar issues arise regarding the way AR relates to the disability movement. Sunaura Taylor notes that the disability community has "largely been left out" of the AR movement, partly because of the movements' "obsession with health and physical fitness," and partly because of a "lack of attention to who has access to different kinds of educational and activist events."[39] If we want what she calls "table fellowship" with the disability movement, we need to be conscious of these exclusionary dimensions of AR activism. We also need to rethink the way we discuss disability in our moral arguments for AR. References to disability have regularly appeared in the AR literature, but almost never in the form of seeking to understand the struggles and aspirations of people with disabilities, let alone seeking to advance those claims. Rather, as disability scholars have noted, AR theorists have often engaged in a kind of "conceptual exploitation" of disability, using people with cognitive disability as a passing (and often ill-informed) thought experiment "to bolster the case" for animals.[40] For example, various AR theorists have invoked the so-called argument from marginal cases (also called the "argument from species overlap") to defend the moral status of animals. This argument assumes that neurotypical adult humans with capacities for linguistic agency have unquestioned moral status—they are the core case of moral status—and that insofar as both animals and people with cognitive disability lack full possession of the relevant capacities underpinning moral status, they both constitute "marginal" cases. The burden of the "argument from marginal cases" is to call for consistency in the way we deal with these so-called marginal cases, so that whatever moral status we accord to "deficient" or "unfortunate" people with cognitive disability, we should do also for animals with comparable cognitive capacities. (And if some animals exhibit cognitive capacities that some people with cognitive disability lack, then, perhaps, these animals should be seen as closer to the neurotypical norm than some humans, and hence accorded higher moral status.)

We have elsewhere criticized the philosophical problems with this argumentative strategy. We entirely reject the premise that moral status is based on a hierarchy of cognitive capacities, or on scales of neurotypicality versus deficiency/deviance. In our view, all beings who have a subjective experience of the world are self-originating sources of moral claims, and hence rights-bearing subjects, regardless of their proximity or distance from any alleged norm of human neurotypicality, and of their place on any alleged scale of cognitive complexity.[41] The argument for AR would not be touched one iota if it turned out that all humans were identical in their cognitive and linguistic abilities, so that there were no "marginal" cases to appeal to for consistency.

But our concern here is with the political flaws of this argumentative strategy, and its pernicious effects on the possibility of progressive alliances. As disability advocates have noted,[42] the "marginal cases" argument not only instrumentalizes disability to "bolster the case" but also perpetuates a deeply problematic conception of neurotypical human cognition as defining the core of moral status, and treats other forms of subjectivity as somehow deficient bases of moral status. Deviation from the norm is arrogantly conceived as misfortune, impairment, deficiency, or temporary embarrassment. Rather than challenging this unwarranted privileging of neurotypical human adults, the appeal to "marginal cases" by AR theorists reinscribes it, and generates perverse comparisons—a kind of jockeying for position in a zero-sum game regarding who falls closest to the privileged norm.[43]

AR theorists did not introduce the idea that proximity or distance from neurotypical adult persons defines degrees of moral status—this idea has a long history in philosophy and bioethics, predating the rise of AR. Unfortunately, however, many AR theorists have failed to recognize that the appropriate response to these arguments is to refuse to accept the starting premise, with its implicit ableism and ageism. AR theorists have thereby alienated disability scholars and activists, when they should be working on a common strategy to combat ableism.[44]

A third example concerns the AR movement's relationship to the struggles of racialized minorities and indigenous peoples. The AR movement has sometimes been overly focused on minority or foreign animal practices, for example, the live export campaign in Australia, which focuses on how Australian animals are subject to cruel slaughter practices in Indonesia, European campaigns against halal and kosher slaughter practices, or North American campaigns against indigenous whale or seal hunts. The unintended result is to reinforce cultural chauvinism and neocolonialism, feeding into popular stereotypes that minorities and foreigners are backward, even barbaric, and sometimes even feeding into explicitly xenophobic and racist anti-immigrant or anti-indigenous political movements. The net result is to alienate potential allies from minority ethnic and racial communities.

To be clear, we are not suggesting that subaltern human groups get a free pass when it comes to their treatment of animals. A commitment to AR entails that all of us can legitimately be asked to account for our treatment of animals. Many AR groups have been quite careful to avoid using stereotypical language or images in their advocacy around minority or foreign animal practices. But they have been less careful in thinking about how their advocacy will be used and abused in the larger political setting, and how it may (unintentionally) operate to reproduce racial hierarchies and to render invisible majority abuses. In contexts where racial and religious hierarchies are pronounced, focusing on minority practices may actually end up reinforcing the apparatus of the animal industrial complex by setting up a chasm between majority (which are perceived as "humane") and minority (perceived as "cruel") culture practices. The result is not just that we lose potential allies in minority communities, and that we may exacerbate patterns of racialization and xenophobia, but also that we may perversely end up strengthening the sense among the majority that they are "civilized" and "humane" in their treatment of animals. AR advocacy should, therefore, target mainstream and majority practices, which exploit animals (animal agriculture, fishing/aquaculture, animal research and entertainment, development and resource extraction industries), and resist media efforts to selectively target minority animal practices as somehow unusually "cruel" or "unnecessary."[45]

In all of these ways, the AR movement needs to think more carefully about how the arguments and campaigns we use in pursuit of justice for animals may unintentionally alienate other social justice movements. If we want to be members in good standing of the family of social justice movements, we need to be responsive and accountable to those movements for our effects on them (just as they need to attend to their impact on the AR movement).[46]

But we would go further and argue that if the AR movement wants to be seen as a member of the family of social justice movements, it may need to rethink the very way it defines justice for animals. At the moment, some of the most influential strands of the AR movement define animal justice in a way that makes it appear to have nothing in common with human justice, and hence as having no relevance for other social justice movements.

For example, starting with Peter Singer's *Animal Liberation*, the AR movement has often adopted what we call a "parsimonious" approach to arguing for the rights of animals, relying on the most minimal assumptions about animals and about human-animal relations. The approach is parsimonious in the sense that it focuses only on animals' capacity to suffer, not their capacities to lead complex and flourishing lives. Animals typically appear as passive victims who suffer at human hands, not as willful agents capable of deciding, of resisting, and of pursuing their subjective good, including how they want to relate to humans.[47] It is also parsimonious with respect to human moral obligations to animals. AR typically emphasizes our negative duty to not cause suffering, rather than positive obligations to attend to animals' well-being or to provide opportunities for them to pursue their flourishing.

There is an obvious strategic rationale for adopting this parsimonious strategy. The hope was that if AR relied on the most minimal premises—both about animals' capacities and about our moral obligations—it would be difficult for people to contest the normative logic of AR. After all, who can deny that animals suffer, and that causing unnecessary suffering is morally wrong? Appealing to more ambitious scientific premises about animals' capacities for agency and flourishing, or to more ambitious moral premises about positive duties to animals, would make it easier to reject AR on the grounds that we are making controversial and unproven scientific and moral claims.

While this parsimonious strategy has a certain logic, the result has been to create an enormous gulf between the fight for AR (conceived in narrow terms as a respect for negative rights, or noninterference, or prevention of unnecessary suffering and cruelty), and the fight for social justice among humans. The latter is always conceived in much more capacious terms, responding to demands for justice and inclusion, and respecting individuals as active, intentional, responsive beings, not just as passive recipients of the duties of others. The parsimonious AR framing has exaggerated (and, indeed, falsely created) differences between animal and human justice, rather than showing how they draw from a common well.

This wedge has been exacerbated by the influence of abolitionist-extinctionist views in the AR movement, which claim that the appropriate response to the injustices of animal domestication is to bring about the extinction of domesticated animals. According to this view, domesticated animals, who have been bred for dependency on us, are incapable of leading decent lives because their dependency makes them inherently degraded and unnatural, and/or because humans cannot be trusted not to exploit their dependency for their self-interested benefits. For one or both of these reasons, extinctionists argue that justice requires domesticated animals to become extinct. We should therefore sterilize (and/or sexually segregate) all existing domesticated animals until they die out.[48] We have elsewhere criticized the inherent perversity of this view, which envisions ending injustice by eliminating its victims, and by using a process (forced sterilization), which would be considered an obvious violation of rights (sexual rights, rights to form a family, rights to bodily integrity) in the human case.[49] But for the purposes of this chapter, the more relevant problem is that it once again makes it seem that the struggle for AR and human rights must be entirely separate matters, operating on different fundamental premises, rather than being continuous in their moral commitments.[50] In no possible case of human social justice would we think that justice is promoted by rendering human victims extinct.

In these and other ways, the very definition of justice within the AR movement has reproduced, and to some extent exacerbated, the status of AR as an orphan of the left. If AR is to be accepted as a valued member of the family of social justice movements, we need to highlight the continuities, and not just the distinctions, with human justice. If the good of animals is continuous with the good of humans, given our shared existential condition as vulnerable embodied subjects, then so too animal justice must be continuous with human justice. Justice toward animals, like justice toward humans, must include obligations to be responsive to their subjectivity and protective of their vulnerabilities. This is the goal of our Zoopolitical theory, linking AR to fundamental social justice principles of membership, participation, citizenship, recognition, and inclusion. Thinking of animal justice in this way, we hope, can help to overcome the alienation of AR, and enable it to join the wider community of social justice movements.

Defining an Inclusive AR Movement

Current AR movement strategy has not only failed to reach out to other social justice movements, it has also, perhaps surprisingly, failed to reach out to its own natural constituency of animal lovers, reformers, and activists. A significant number of people who devote their lives to the cause of animals, and who are in practice upholding fundamental claims of justice for animals, do not identify with the current AR movement. Again, this is partly due to the success of governmental and industry efforts to tar AR with the extremist label. But it is also partly due to the way the movement has defined its own "moral baseline" for membership and acceptance.

The idea that a social justice movement requires a moral baseline is understandable. As soon as a social justice movement achieves any level of success or influence, its language and symbols can quickly become co-opted by the powers that be. We can see this with the feminist, gay, and environmental movements, and the resulting "pinkwashing" and "greenwashing" of corporate interests, and also with animal advocacy, as various actors in the animal industrial complex rebrand themselves as "humane," "caring," and "animal-friendly," even as they continue or even intensify their violent exploitation of animals.

It is natural, in this context, to want to identify some moral baseline that allows us to distinguish those whose commitment to animals is merely token or rhetorical from those who truly prioritize animal interests and AR. But what is this baseline? What is the minimum moral commitment that is needed to be considered a legitimate participant in the AR movement?

For many people, this moral baseline is an individual ethical commitment to not directly participate in animal exploitation, which in turn is captured by a commitment to veganism.[51] This view that veganism is the moral baseline has been described by Ian Smith as follows:

> The "veganism as a moral baseline" idea is most commonly used to demarcate "us" and "them"; it goes beyond the claim that veganism is praiseworthy or even morally obligatory and posits that veganism is *the* litmus test for credibility and participation within the movement. Deviating from a vegan diet (or perhaps more accurately a vegan lifestyle) cannot be compensated for with other actions.[52]

In our view, this emphasis on veganism as a moral baseline is strategically misplaced. To be sure, one of the most important steps a person can take is to end their personal direct exploitation of animals by becoming vegan, and there will be no justice for animals until we renounce the right to eat them. But there are many other requirements of justice for animals, and many other ways to fight for AR: protecting habitat from development or pollution; fighting climate change; rescuing and rehabilitating injured wild animals; developing animal-friendly building codes and technologies; banning marine parks, zoos, and circuses that use wild animals; fighting animal agricultural subsidies and requiring the industry to pay all externalized costs; demanding adoption of alternatives to animal research; promoting co-existence practices with urban wildlife; exposing animal exploitation and educating others about it; rescuing and caring for abandoned animals; fighting for legal rights for animals; fighting for political representation for animals; and so on. Raffaella Ciavatta captures the point succinctly with her slogan "Activism is my moral baseline."[53]

If we say that veganism is the moral baseline, then all of the people who are working for AR in the ways just noted, but who are not vegan, are told they are not part of the movement. They have failed the entrance requirement. All of the work they do for animals (perhaps caring 24/7 for orphaned birds, then releasing them to the wild; or tirelessly lobbying city hall to adopt animal-friendly zoning and building codes) counts less than, say, an individual who spends his days as a banker, lawyer, carbon industry worker, institutional food buyer, health inspector, or any of countless other occupations which advance corporate or state agendas that harm animals in multiple ways, but who is an "ethical" vegan and, therefore, qualifies as meeting the moral baseline.[54]

This individual ethical conception of AR is deeply limiting for the movement. As noted earlier, individual behavior is subject to significant backsliding. And there is a serious question about how effective consumer boycott (in isolation) is as a long-term strategy for bringing about institutional change.[55] Veganism is a difficult personal challenge for many people in the context of current institutions and environments. They need a supportive environment in place before they can change their individual behavior. They cannot "go it alone," perhaps, because they are not willing to be socially awkward, or deviant, or to be perceived as morally sanctimonious, or because they simply cannot muster the necessary energy to act against convention. Moreover, veganism is subject to collective action problems. As noted earlier, when it comes to ratcheting up our personal moral behavior, many people express a theoretical willingness to do so, but not under conditions in which they feel they are making a sacrifice which others are not.[56] So they may support regulatory change that requires everyone to alter their behavior, before they will change their individual behavior. A further problem is what social psychologists call moral self-licensing. When we do something we perceive as good, in the sense of going beyond what is expected or required by existing social norms, we tend to license ourselves to do something bad, reestablishing a kind of moral equilibrium in relation to social norms[57] (e.g. if you eat vegan, you might then feel licensed to drive an SUV, or buy sweat-shopped clothes). These awkward challenges to individual moral action underline the importance of shifting social norms, thus raising the moral tide that we're all swimming in.

This connects to a larger concern about perceptions of AR as a private morality movement, not a public policy movement. According to Norm Phelps, the general public perceives AR as part of the same regressive camp as prohibition, the war on drugs and the anti-abortion movement, rather than as part of the social justice left (civil rights, feminism, LGBT, etc.), "in large part because we place so much emphasis on personal dietary decisions and comparatively little emphasis on institutional and societal attitudes toward animals."[58] This may be an unfair perception on the part of the general public, but it is fueled every time we put a single-minded emphasis on veganism, rather than broad-based structural and institutional changes across a range of issues and locations, and in alliance with other progressives.

None of this is to say that promoting veganism is not a crucial part of AR activism. We cannot achieve justice for animals if we continue to assert a right to eat them—this is a central dimension of the unthinking "habitus" of human violence toward animals which must be disrupted. But so too we cannot achieve justice for animals if we continue to colonize their habitat, degrade their water supplies, build roads, fences, and wires that are lethal to them, confine them, or abandon them to neglect. We can hope and expect that participating in the AR movement will make everyone more conscious of all of the ways we negatively affect the lives of animals. But it is a mistake to present any one of these as a baseline, that is, as the prerequisite step to be part of the AR movement. There are many people working on discrete animal campaigns (e.g. companion animal access rights, coyote co-existence programs, farm animal sanctuary or welfare reform), or as part of the environmental movement, or the indigenous rights movement, who are fighting for AR (rights to life, liberty, protection, and care; rights to land and habitat) and who are not necessarily (or not yet!) vegans.[59] Each in his or her own way is denaturalizing human violence against animals, and reshaping the habitus to create more just human-animal relations. We must not erect entrance barriers to them being part of the AR movement, or treat them like second-class activists and allies.

In short, the contemporary AR movement has defined its arguments, tactics, and self-identity in ways that tend to contract, rather than grow, the movement. The arguments and tactics do not operate to form bridges with other social justice movements, or even with the full range of activists and reformers who work to defend the rights of animals. We believe that it is one virtue of our Zoopolitical theory of AR that it can overcome these barriers to a more successful movement. By shifting the emphasis from personal ethics to politics, and by drawing inspiration from progressive social movements of the postwar era, we can situate AR within the larger family of social justice struggles for inclusive and participatory citizenship. It's not just that there are lessons to be learned from other social progressives; it's that we need to be part of the broader social justice movement with them, and this requires reframing AR theory and advocacy in ways that build on our common foundations, rather than creating unnecessary cleavages.

SOCIAL JUSTICE FROM THE GROUND UP

So far we have argued that a broadly liberal approach to AR can support genuinely transformative change, but to do so, it must connect AR to the ideals and narratives that underpin broader social justice struggles, and it must focus not only on the "cognitive subversion" of ideological beliefs but also on changing people's everyday lived environments. We've argued that this requires significant reorientation in how we frame AR arguments, how we define AR advocacy, and how we create the conditions for alliance politics.

This is all still rather abstract, and we will conclude the chapter by offering seven more specific proposals for how to advance an AR agenda. But we might first note the implications of our analysis for Cavalieri's own proposals regarding the struggle for the legal recognition of personhood for great apes and whales. If our analysis is sound, it suggests certain limits to such a strategy. As we noted earlier, achieving legal personhood for even just one nonhuman species might, indeed, stand as a powerful disruption to inherited ideologies of human supremacism. But it is less clear that it helps forge links with other social justice movements, in part because these particular species are largely absent from our everyday lives. Very few people have any direct relationship to great apes or whales: they are not part of our social context. So, whatever changes follow from according personhood to great apes or whales, they are unlikely to change anything in our everyday lived experience. Someone could support personhood for these beings, and then never have a single interaction with an ape or whale for the rest of his or her life. And so we might worry that even if a country endorsed personhood for great apes, it would do little if anything to change people's expectations and habits regarding the animals that actually are part of their everyday lives. New Zealand will be an interesting test case in this regard, having (partially) endorsed basic rights for great apes, while still operating an agricultural economy that is entirely built on the exploitation of domesticated animals. Personhood for great apes does nothing, by itself, to change our views of how domesticated animals fit into our society, what sort of social role they play, and what kinds of membership they should be accorded. At the end of the day, it does not really affect the "habitus" in a deep, lived sense.

Our own thinking starts from a different location. We are interested in how communities can be transformed from the ground up—undertaking the kind of self-organization and institutional and structural reform that can nourish and sustain the fight for social justice over the long haul. We believe that transforming communities where people live—the meaningful localities of everyday life, and the spaces and places where citizenship is actually enacted—is needed to initiate the intergenerational iterative process of norm change. As our example of Stan suggests, when experience living in a different environment which supports different ways of being in relation to animals, their behavior and choices will change, which, in turn, will support the development of new institutions and new norms.

We are not proposing that AR movement strategy be restricted to grassroots politics and community-building. On the contrary, we strongly support coordinated advocacy for legal and policy changes at multiple levels (referenda, lobbying and party politics; litigation and constitutional challenges; national and international campaigns, etc.). This is essential given our earlier comments that the AR movement is too often perceived as a movement about private morality or lifestyle choice rather than public justice. These policy-making processes are also essential sites for the reproduction of the broader social imaginary, with its norms and practices of democracy, citizenship, shared community, and social justice. As we argued earlier, the AR movement needs to be drawing upon, and engaging with, this social imaginary. This, in turn, requires careful strategic thinking about AR's relationship with political parties, who, for better or worse, remain the central vehicles for defining public agendas. As Kim Stallwood notes, the lack of engagement with political parties is one of the most striking failures of the AR movement, at least in Britain and the USA.[60]

So there is much work to be done in terms of developing coherent advocacy strategies at the level of national (and international) politics. But these strategies will inevitably vary from country to country, depending on the electoral system (e.g. does a country have a proportional representation system open to new parties?),[61] the scope for referenda or other popular initiatives (e.g. can citizens put an animal issue on the ballot?), and the scope of possible litigation (e.g. do citizens have standing to bring suits on behalf of animals?). These details are essential to any sensible political or legal strategy, but vary from country to country in a way that makes it impossible to generalize.

In any event, our focus in this section is on a parallel process of grass-roots politics and community-building, which we believe needs to accompany national legislative or litigation strategies. Grassroots politics and community-building are crucial dimensions of building a successful AR movement, for, at least, two reasons. First, as noted earlier, progressive social justice movements depend on an "infrastructure of dissent," tied to the social texture of people's everyday lives, within which people have the safety and support needed to sustain unpopular positions. This is true of all justice movements. But, in the case of AR, there is a second reason to highlight community-building: it is only through building specific sorts of local spaces that we can even identify what justice requires in relation to animals. We cannot know what interspecies justice requires without learning what sorts of relationships animals (as individuals and social groups) want to have with us, and that, in turn, requires creating conditions that enable animals to express their preferences about the kinds of lives they want to lead.[62] We must, in short, enable animals to become agents in defining their relationship to us, and this is a task best suited to local communities. In the spaces and places of our day-to-day lives, as opposed to more structured, detached, virtual, or language-mediated contexts, we can observe and get to know individual animal others. We can engage with them in contexts where they can make meaningful decisions about their lives, learning with, and from them, about the possibilities of interspecies community.[63]

We will briefly discuss seven clusters in which the AR movement can advance the process of norm change concerning human-animal relations in ways that (1) form alliances with other social progressives, and (2) create spaces for animals to be agents in defining their relations with us, namely, the new commons; inclusive design; the interdependent/caring community; green cities and corridors; a progressive place-based economy; universities/education; and the grassroots democratic reform movement.

The New Commons

A crucial dimension of the fight for (human) economic justice concerns the right to place, or to housing—a place to call home. Neoliberal economics and the marketization of most land and housing put continuous pressure on property values and create chronic insecurity for the poor who are denied the most basic of social rights. Progressives have developed various strategies, from squatting in abandoned buildings, to reclaiming foreclosed properties, and to creating community land trusts in order to challenge neoliberal economics, with some success.[64] Related movements, such as community gardening, "guerilla" gardening, and food security activism, claim underused or misused property or public spaces in order to support the basic needs of the local human population.

Consider, now, the situation of animals in the city, whether domesticated or liminal (i.e. nondomesticated urban dwellers, such as raccoons, peregrine falcons, foxes, groundhogs, or sparrows) and the fundamental ways in which they are denied a right to space and a home. Farmed animals are zoned into smaller and smaller containers, animal companions are shut up in crates and runs, and liminal animals are treated as pests and interlopers wherever they are. The fight for justice for animals starts with the battle to belong, to be viewed as members of the community, with the right to secure space.

If we bring these two sets of issues together, we can see how a movement to create community land trusts or a new "commons" could integrate and advance both human and animal social justice—an interspecies commons where everyone has a home, and food security. For example, a land trust model could reclaim urban brown lands, abandoned properties, or misused public space, and redevelop them to combine co-op housing, a micro-sanctuary for rescued farmed animals and feral cats,[65] veganic community gardens, and green spaces to accommodate liminal animals. Over time, following models established in Burlington, Vermont, and elsewhere, such land trusts can gradually expand, reclaiming a certain percentage of land from private markets and rededicating it to communal and public ends under direct democratic control—a commons which recognizes the right to basic housing and food security for all members of the community.[66]

Inclusive Design

Another key dimension of the struggle for human social justice centers on issues of inclusion and access. Since the 1960s, the universal design movement has worked toward creating public buildings and spaces, which are deliberately designed to be accessible to the widest population possible (old people, young people, people with disabilities), rather than simply defaulting to an unexamined ableist adult norm. Now, imagine if we were to integrate domesticated and liminal animals into this idea of universal/inclusive design. The unexamined default norm is that cities are human spaces, and that design should only take their interests and needs into account. This is, of course, unfair to animals, relegating them to the status of barely tolerated, invisible, stigmatized, or unwelcome members of the community. It also reproduces human injustices since human and animal mobility and access are closely intertwined. For example, when companion animals are unwelcome in public spaces, businesses, and on public transportation, the mobility of their human companions is restricted too. When it comes to liminal animals, failure to incorporate their needs into city planning creates needless problems for humans, who have to deal with animals getting into the attic, or the garbage, or fouling the beach, or darting across the road, or starting a fire in the chimney, or smashing into windows. Many of these conflicts could be resolved by practicing interspecies universal design, which takes into account the needs and habits of liminal animals, namely, their mobility requirements and access to water, food, shelter, and safety. As we have argued elsewhere, this does not mean laying down the red carpet for liminal animals, but rather planning better to avoid needless and avoidable dangers and conflicts.[67]

The Interdependent/Caring Community

For decades, feminist theorists have demonstrated how traditional theories of justice obscure the reality that humans are interdependent beings, not self-sufficient individuals. The myth of independence has been constructed via a public/private distinction which relegates the vast majority of work traditionally performed by women (reproductive work, domestic work, caring work) to the private sphere, ignoring the ways in which men's (traditional) work in the public sphere has been dependent on this invisible labor. Until recently, issues of justice concerning work in the so-called private sphere—recognition, compensation, distribution, working conditions, choice, and opportunity—were completely ignored. And liberal democracies are still a long way from adequately recognizing the realities of care/reproductive work, or from rethinking a public/private distinction which continues to stand in the way of justice. The overwhelming burden of this work continues to be borne by individual women in families, and by grossly underpaid women working as childcare workers, elder and community care providers, cleaners, food service workers, social and public health workers, and so on.

Interestingly, another form of care work undertaken overwhelmingly by women is caring for animals as sanctuary and rescue workers, wildlife rehabbers, companion animal service providers, or vets. Animals, in turn, are performing ever-increasing amounts of care work as companions, hospital visitors, therapy friends, and service animals for people with disabilities.[68] Much of this work is invisible, and many of these care workers are exploited.

Care work is largely performed either within the family home, where the imbalance in who performs it is invisible, or in wage ghettos and total institutions. The problem is not simply that the work is under recognized and undercompensated. The deeper problem is that modern economies—having been built around the myth of the autonomous wage earner, whose care needs are conveniently attended to behind the veil of the private sphere—must be completely restructured to accommodate an interdependent work force. One approach to reorganizing care work is to develop integrated residence-work-care models. Instead of caring work being located primarily in either isolated homes or total institutions (animal sanctuaries, seniors homes, daycares, residential institutions for people with disabilities, correctional facilities, etc.), which segregate a special-needs population from the rest of the community, care work would be integrated and distributed across the settings of daily life.

For example, there are already working models of correctional facilities which integrate programs for socialization and adoption of abandoned dogs and cats. Offenders benefit from learning to love and care for the animals; the dogs and cats benefit from this friendship, and the chance to eventually become part of a family. This model could be expanded to incorporate wild animal rescue and rehab, or farmed animal sanctuaries, or veganic farms. The recognition that lonely, bored, and isolated seniors greatly desire the companionship of animals (but feel they cannot fully care for them) has led to the development of interspecies and intergenerational supported living communities for seniors.[69] In such communities, seniors, rather than being simply a cared-for population can also make a contribution to the lives of animals (and children, where daycare and schools are part of the model). Imagine what would happen if, for example, companion animal daycares, or wildlife rehab centers, or liminal animal conservation efforts, were physically integrated with seniors' residential communities (or hospitals, or schools) as a matter of course. Dogs, who would otherwise be cooped up alone in a house all day get to spend time with other dogs and humans. Seniors who want to spend time with animals can sit and observe the animals, and help with feeding, or with other light care activities, or with natural landscaping and habitat construction. The explosion of animal-assisted therapy (AAT) demonstrates the growing realization that humans' mental and physical health benefits from being part of community with animals. Currently, AAT operates (largely) to exploit animals for this purpose as part of piecemeal therapy programs. But a different model is possible, with similar benefits for human well-being, in which animals are not used as tools for human benefit, but are part of a mutual society oriented to ensuring everyone's flourishing.[70]

Integrated residence-work-school-care models offer a way of dismantling public-private divisions, and care provider/care recipient divisions, providing the kind of flexibility to make it easier for everyone to contribute according to their abilities, and receive according to their needs.

Green Cities and Corridors

It is abundantly clear that humans must transition from a carbon-based growth economy. From an environmental perspective, among the many changes necessary will be moving to a plant-based diet; switching to public transportation, conservation, and green energy; reducing human population growth; and protecting and expanding forests, wetlands, and other vital ecologies. All of these changes will also benefit wild and liminal animals, and so animal advocates and environmental activists should be able to find common cause across a range of issues. At the grassroots level, this common cause is, perhaps, most important in the area of transportation and development planning. Urban sprawl, road development, and expanses of hard surface are devastating for liminal and wild animals. They need habitat, green corridors, and most of all they need for humans to stop encroaching on the lands where they live. When and where development does occur, it needs to actively and mindfully attend to the needs of animals for road under/overpasses; access routes to water, food, and shelter; and animal-friendly building construction.[71] These represent crucial dimensions of AR activism, and opportunities to reshape the lived environment in ways that alter the way people relate to other animals.

A Progressive, Place-Based Economy

What does it mean to develop an animal-friendly economy? If one looks at some popular AR Web sites, one might think this involves working for an animal advocacy organization that acts like a typical corporation, or becoming a vegan "entrepreneur" and capitalizing on the next great vegan product. But the AR movement does not have to embrace these models, or at any rate does not have to be restricted to them. We can, instead, establish animal-friendly organizations and businesses, which actively resist neoliberalism, rather than accommodating its dictates. Instead of surrendering to the inevitability of growth and capital mobility, animal-friendly businesses could deliberately choose to be place-dependent anchors for local communities, providing secure employment, and a predictable presence and involvement in the community. Some animal-friendly organizations and businesses already function this way—anchoring themselves in a local community where they develop business and social connections, and a reputation as community builders (promoting local cultural events, supporting social justice endeavors, participating in public health campaigns). In this way, they earn respect and recognition for the AR movement. Over time, they can influence those around to see the opportunities in developing an economy around a just interspecies community (e.g. the economic potential of plant-based farming and vegan cuisine; recreational and tourism possibilities premised on mutually beneficial human-animal relations rather than hunting, trapping, and fishing; or the business opportunities of developing expertise on co-existence and fair accommodation with liminal animals rather than "pest" extermination).

Animal-friendly (or potentially friendly) businesses and organizations are growing at an exponential rate, for example, veganic farming, food production, and restaurants; producers and sellers of vegan clothing and accessories; liminal animal co-existence companies; wildlife protection and habitat conservation organizations; animal companion care and vet service providers; farmed animal sanctuaries and wildlife rehabbers; eco-sensitive tourism and recreation providers; citizen science groups; educators, creators, and religious groups who draw inspiration from connections with animals; animal advocacy organizations.

Each one of these businesses/organizations represents a chance to push back against neoliberal economics, and the social injustices built into models of endless growth, externalized costs, and exploitation of labor. Every animal-friendly business is an opportunity to implement diverse hiring, transparent and ethical business practices for product sourcing and trade, ecological responsibility, community contribution, progressive models of worker management/ownership, and so on. In other words, we can either develop animal-friendly businesses that remain isolated and marginal in the take-no-prisoners game of neoliberalism, or we can consciously embrace a broader progressive agenda in how we do business, and thereby significantly increase our chances of being part of a larger, and more effective, movement for social justice.

Universities/Education

As Anderson notes, a key stage in the development of a social movement, and its power to influence social norms, is the emergence of influential moral leaders, as well as theorists to provide credibility, and to articulate ideals that counter the power of the reigning ideological habitus.[72] Frequently, when "animal issues" are covered in the news (e.g. exposés of factory farms, controversies over animal culls, or civil disobedience by anti-vivisectionists), the "sides" represented by the media will include spokespersons for industry on the one side, and animal activists on the other. Glaringly absent are the voices of researchers and academics to place the issues in context and provide authoritative alternatives to standard ways of understanding and framing controversies. Where are the lawyers, political scientists, and policy experts who can speak about AR issues in this way? They are shockingly thin on the ground, and this is a real disadvantage to AR activists, who need this third voice to add gravitas to the issues raised by their campaigns.

The AR movement has made significant progress in the last decade in terms of institutionalizing critical animal studies in universities, and this should continue to be a priority, especially in those disciplines (e.g. politics, law, sociology, economics) with close links to governance and public policy. For decades, even centuries, theorists in the social sciences have assumed that society is a human phenomenon—that basic issues about justice and sociability are relevant only to interhuman relations. But the social sciences are experiencing an "animal turn," and as this takes hold—that is, as these disciplines begin to appreciate that we live in multispecies societies, not strictly human ones—the vast potential of the acquired knowledge of the social science disciplines can be turned to understanding human-animal relations. It would be hard to overestimate the potential impact of this shift, and the resources it can provide to the AR movement.

A much broader education agenda concerns the systematic socialization of children into the habitus of human exceptionalism. A massive educational enterprise is dedicated to shaping the unconscious and unexamined acceptance of human violence toward animals which characterizes typical human adulthood.[73] This is particularly visible in the life sciences curriculum, which takes children's fascination with, and love for, animals, and over the course of primary, secondary, and tertiary education replaces this with a reductionist and desensitized attitude toward animals as objects of invasive study, rather than the companions, friends, and fellow creatures of infancy and childhood. Reaching out to teachers and curriculum developers about the possibilities for a genuinely inclusive and interspecies model of education is a crucial dimension of animal advocacy.

Grassroots Democratic Reform

It is no secret that democracies are struggling. The political process, especially at national and supra-national levels, has been significantly captured by special interests. Traditional forms of political participation (party membership, voting) are plummeting. While new social media create openings for different kinds of political activism and participation, they are also powerful weapons in the hands of those who wish to monitor and control dissent. In light of these challenges, grassroots democratic reform focuses on reenergizing democracies at the local and regional levels, promoting opportunities for citizens to participate directly in decision-making about issues that touch directly on their daily lives. Many of these issues, as it turns out, concern animals. In many countries, local and regional governments still retain significant power in determining policies that directly affect animals—from zoning and development policies, to animal "control," to issuing licenses for companion animals, to food industry regulations. Many local jurisdictions have imposed bans (e.g. foie gras, circuses, carriage horses, pet sales, exotic pets, dog racing) and zoning controls (e.g. on factory farms, puppy mills). Local governments can also act to protect habitat and green corridors; they can implement building codes, exercise considerable purchasing power (e.g. food sourcing), and adopt co-existence, rather than extermination policies for dealing with liminal animals.[74]

Given the range of local governance issues concerning animals, it makes sense for the AR movement to focus some efforts at this level to institutionalize representation of animals in decision-making. Rather than issues of dog parks, or farm zoning, or wild animal culls being framed as conflicts between groups of humans, it is crucial to insert direct representation of animals' interests into the equation. This could be in the form of an animal ombudsperson or advocate on city council, for example, with genuine power to affect decision-making. Eventually, the goal should be to have advocates in all key sectors—on planning boards, hospital boards, social service boards, school boards, business associations, universities, and so on—someone who is tasked with representing the animals whose interests are affected by decision-making. Similar initiatives are in place to represent others who cannot elect representatives, such as young children and people with cognitive disabilities. In recent decades, the children's rights and disability rights movements have made a concerted push to insist that these groups "have a say in matters affecting them," and as we have argued elsewhere, the AR movement can benefit directly from the models of "dependent agency" that have been developed.[75]

The AR movement has been actively targeted by anti-terrorism legislation, ag-gag laws, and concerted efforts to marginalize the movement as extremist. We have vital common cause here with civil libertarians and democracy activists, if we make the effort to frame our concerns in ways that underline the relevant commonalities. For example, in our own country, Canada, it would be hard to imagine a practice more secretive and undemocratic than decision-making around the use of animals in biomedical research. Much of this research takes place in universities full of democratic theorists and governance experts who have never given the issue a second thought, even though the alleged mechanisms to monitor animal research are a paradigm of regulatory capture.[76] Animal agriculture represents a similarly entrenched, and largely unexamined, instance of total regulatory capture which should be of deep concern to anyone who cares about democracy.

As in developing a progressive place-based economy, the struggle for local animal democracy represents an opportunity for the AR movement to approach activism in ways that either enhance, or limit, potential alliances. By choosing to pursue democratic reform for animals in ways that respect transparent and open debate, encourage broad participation, embrace diversity, experiment, and compromise, and think creatively about potential alliances, we have the potential to embed and grow the AR movement in a strong broad-based movement for democratic renewal and social justice.

CONCLUSION

Our ability to envision an effective AR movement has been hampered by a series of inadequate and inappropriate frameworks. These include the following:

1. totalizing explanations, which suggest that the AR cause will be lost or won depending on our successful mastery of sweeping historical logics, whether it is the logic of capitalism, enlightenment rationality, humanism, or the rights revolution, rather than more situated and contingent relations and strategies;
2. a naïve faith in the power of rational argument and ideological critique to sway human beliefs and behaviors that are, in fact, strongly shaped by identity, implicit knowledge, and habit;
3. political strategies focused primarily on the role of individual conscience and choice rather than collective organization and a struggle for power;
4. a definition of the AR movement in terms of "moral baselines," which separate us from our natural allies, rather than providing a basis for forging connections and a broad-based movement for social justice.

Instead, we have argued that there is no avoiding the long, hard road of engaged politics and grassroots community and institution building—of battles fought over and over in countless locales over many decades, of the inevitable setbacks, and the lack of certainty about whether we are making progress, and whether it will stick. We believe that history provides many examples of how this kind of change has been achieved in liberal democracies when sufficient numbers of people are motivated and organized to fight for a cause over the long haul.

To do this, we need to build a much bigger movement, by galvanizing, rather than alienating, our natural allies. We need to start by changing the world where people live, one community at a time, and thereby altering their inherited habitus of human-animal hierarchy. As John Sanbonmatsu says, "We want to build, here and now in our movement, in a concrete way, a mini-version of the idealized society of the future that we are striving towards."[77] It is by giving people a real taste and a positive vision of this possible interspecies world of social justice that we can achieve the strength, and the numbers, to challenge the reigning paradigm of total exploitation of nonhuman animals.

NOTES

1. The liberal tradition has multiple strands, many of which are, indeed, inadequate to the needs of the AR movement (or of other social justice movements). Our liberal approach is rooted in left liberal (or liberal egalitarian) conceptions of justice (offering a strong critique of the institutions of existing liberal democracies and capitalist political economies), as against right liberal or libertarian defenses of market inequalities and market fundamentalism; multicultural conceptions of citizenship (which recognize group-differentiated forms of membership), as against universalistic or difference-blind conceptions of citizenship; bounded conceptions of political community (which recognize legitimate interests in collective self-government), as against (non-rooted) cosmopolitan conceptions that dissolve any meaningful role for states or other bounded political communities.

    In our previous work, we have argued that this version of liberal citizenship—egalitarian, bounded, and diverse—provides a compelling philosophical foundation for a multicultural Zoopolis. Needless to say, all of these claims are controversial within the liberal tradition. For a defense of this interpretation of liberalism, and critiques of rival interpretations, see Will Kymlicka, *Contemporary Political Philosophy: An Introduction* (Oxford: Oxford University Press, 2002), particularly Chaps. 3 and 7.

2. Paola Cavalieri, "Animal Liberation: A Political Perspective," in this volume, 19. Subsequent references to her chapter will be cited in parentheses in the text.

3. For critiques of vegan strategizing that emphasizes individual choice over broader social justice action, see Wayne Hsiung, "Boycott Veganism," *Direct Action Everywhere*, http://files.meetup.com/160880/Boycott%20 veganism.pdf; Jon Hochschartner, "Should a Meat-Eater Advocate for a Vegan Society?," *CounterPunch*, March 7, 2014, http://www.counter-punch.org/2014/03/07/should-a-meat-eater-advocate-for-a-vegan-society/; Norm Phelps, *Changing the Game: Why the Battle for Animal Liberation is So Hard and How We Can Win It* (revised edition, Herndon, VA: Lantern, 2015); Collectively Free collective, "Why Activism and Direct Action?," http://collectivelyfree.org/why-activism-and-direct-action/.

4. Paola Cavalieri, *The Animal Question: Why Nonhuman Animals Deserve Human Rights* (Oxford: Oxford University Press, 2001).

5. Paola Cavalieri and Peter Singer, eds., *The Great Ape Project: Equality Beyond Humanity* (London: Fourth Estate, 1993); Cavalieri, "Whales as persons," in *Ethics and the politics of food*, ed. M. Kaiser and M. Lien (Wageningen: Wageningen Academic Publishers, 2006).

6. Jonathan Haidt, "The New Synthesis in Moral Psychology," *Science* 316 (2007): 998–1002; Lisa Bortolotti, "Does reflection lead to wise choices?,"

Philosophical Explorations 14 (2011): 297–313; Valerie Tiberius and Jason Swartwood, "Wisdom revisited: a case study in normative theorizing," *Philosophical Explorations* 14 (2011): 277–295.

7. Che Green, "How Many Former Vegetarians and Vegans Are There?," Humane Research Council, December 2, 2014, https://faunalytics.org/ how-many-former-vegetarians-and-vegans-are-there/. See also the "Vegetarian Recidivism" page of the Animal Charity Evaluators, Web site at http://www. animalcharityevaluators.org/research/foundational-research/ vegetarian-recidivism/.

8. Hal Herzog, "Why Do Most Vegetarians Go Back To Eating Meat?" (blog), *Psychology Today*, June 20, 2011, http://www.psychologytoday. com/blog/animals-and-us/201106/why-do-most-vegetarians-go-back-eating-meat.

9. As Norm Phelps puts it, given the level of recidivism, the vegan outreach strategy has been "treading water" (Phelps, *Changing the Game*), 1%.

10. Peter Dauvergne and Genevieve Lebaron, *Protest Inc: The Corporatization of Activism* (London: Polity Press, 2014), 17–18, 85. They emphasize that it is very difficult, if not impossible, to sustain this infrastructure of dissent based solely on a shared endorsement of explicit principles, without the sort of social glue that comes from less reflective aspects of our social lives, such as shared experiences, identities, territorial attachments, and ways of life.

11. More exactly, we endorse the view that all sentient animals, human and nonhuman, should be seen as rights-bearing subjects. In *Zoopolis*, we argued that "selfhood," rather than "personhood," might be a more perspicuous term to denote this status, both for humans or animals, given the multiple meanings that "personhood" has had in the law and in philosophy. But we also recognize that, for better or worse, "personhood" has become the legal category used to denote the status of a rights-bearing subject, and insofar as this is so, then the pursuit of basic rights for humans and animals will need to be conducted in the language of personhood (Donaldson and Kymlicka, *Zoopolis: A Political Theory of Animal Rights* [Oxford: Oxford University Press, 2011], 24–32).

12. David Nibert, *Animal rights/human rights: Entanglements of oppression and liberation* (Lanham: Rowman & Littlefield, 2002).

13. Sue Donaldson and Will Kymlicka, "Reply to Svärd, Nurse, and Ryland," *Journal of Animal Ethics* 3/2 (2013): 208–219, and "Reply: Animal Citizenship, Liberal Theory and the Historical Moment," *Dialogue* 52/4 (2013): 769–786; Angus Taylor, "Interview with Sue Donaldson and Will Kymlicka," *Between the Species* 17/1 (2014): 140–165; Dinesh Wadiwel, "Liberalism and Animal Rights: An Interview with Sue Donaldson," Sydney Environment Institute (2014), http://sydney.edu.au/environ-ment-institute/news/sue-donaldson-speaks-to-dinesh-wadiwel/.

14. Our aim in this chapter is to respond to those radical AR advocates who, believing that liberalism lacks any viable strategy for radical change, conclude that we must look beyond liberalism to find a basis for AR advocacy. We should note, however, that some liberals share the perception that our work is wildly utopian, and conclude that liberals must, therefore, be much more modest in their aims. Robert Garner, for example, suggests that Zoopolis is unachievable because it expects too much of humans (the sacrifice of too much self-interest), and is premised on a kind of species-egalitarian vision, which is untenable (Robert Garner, *A theory of justice for animals: Animal rights in a nonideal world* [Oxford: Oxford University Press, 2013], 120). For Garner, humans are incapable of recognizing and respecting nonhumans as fellow creatures, let alone as co-citizens of political communities and co-creators of a shared social life. In short, Zoopolis fails Rawls' test of a realistic utopia, which "takes men [*sic*] as they are, and the laws as they might be." It fails because people "as they are" cannot be transformed in the necessary ways (i.e. humans cannot overcome a fundamental species chauvinism, making a species-egalitarian vision impossible). We should, therefore, settle for a more realistic reform of animal welfare policy and practice, one that is premised on humans treating animals humanely, but not as equals. To use Cavalieri's terms, we might say that Garner falls into the "deficient in aim" camp of liberalism. While our main goal in this chapter is to respond to radical critics, if our account is successful, it should also serve as a response to Garner's critique.

15. For example, Martin O'Neill and Thad Williamson, eds., *Property-Owning Democracy: Rawls and Beyond* (Oxford: Blackwell, 2014).

16. Simone Chambers, "Justice or Legitimacy, Barricades or Public Reason? The Politics of Property-Owning Democracy," in O'Neill and Williamson, *Property-Owning Democracy*, 29.

17. Bonnie Honig, *Emergency Politics: Paradox, Law, Democracy* (Princeton: Princeton University Press, 2009), 67.

18. In ecology, shifting baseline syndrome refers to the idea that each generation takes as "normal" the ecological conditions it grew up with, and measures progress or regress in relation to that baseline, even if, in the longer sweep of history, it reflects a deeply degraded situation (e.g. with regard to species diversity, or pollution levels). Without the longer view, our ideas for the range of possibilities, good and bad, for life on earth can be distorted and impoverished. In the context of struggles for human justice, this short-sighted vision means we fail to see the impact of long-term incremental change. As progressive change occurs and becomes "normalized," we forget that it has been achieved through heroic effort against great odds of power, and against the conservative dynamics of the social habitus. We forget that a century ago, beating children was ubiquitous practice in industrialized

Western societies, considered essential to successful socialization. This practice was not overturned by a violent revolution or "break in the course of history." It happened incrementally and diffusely, generating a new social habitus, in ways that make it hard to believe that very different views and practices once held sway. What was once perceived as a fact of human nature ("the way men are") turns out to have been contingent social practice. See Steven Pinker, *Better Angels of Our Nature* (New York: Penguin, 2011); Hope Babcock, "Assuming Personal Responsibility for Improving the Environment: Moving Toward a New Environmental Norm," *Harvard Environmental Law Review* 33 (2009): 117–175.

19. Gar Alperovitz, "The Pluralist Commonwealth and Property-Owning Democracy," in O'Neill and Williamson, *Property-Owning Democracy*; James DeFilippis, *Unmaking Goliath: Community Control in the Face of Global Capital* (New York, Routledge, 2003).

20. Joe Hoover, "The human right to housing and community empowerment: home occupation, eviction defence and community land trusts," *Third World Quarterly* 36/5 (2015): 16.

21. For this contrast between rights as tools of governance and rights as tools of resistance, see Hoover, "Human Right to Housing." (For a related work that recuperates the resistance function of rights, see Guy Aitchison, "Rights, Citizenship and Political Struggle," *European Journal of Political Theory* [2015].) According to Hoover, when the state fines a landlord for racially discriminatory rental practices, this is an example of rights as "governance," but when homeless people squat in abandoned buildings, or sub-prime mortgage victims reclaim their empty homes from bank repossession, these are examples of rights as "resistance" (ibid., 3). In the latter case, "The women and men fighting for a human right to housing are not using human rights only to seek concessions from the powerful, nor are they using human rights to ask for protection in the language that power recognises, they are claiming rights in their own vernacular, drawing on their own experiences, and attempting to use the idea of human rights to empower themselves and their communities" (ibid., 17). In either mode, rights can be claimed or enforced in ways that advance or undermine justice, so the point is not that state authority and enforcement is bad, and radical challenge is good. The point rather is to recognize that while at this stage in history, vis-à-vis animals, rights operate as tools of governance to dominate animals, this does not negate the power of rights-as-resistance for advancing new claims on behalf of animals. See also Jedediah Purdy's observation regarding the idea of human rights: "While mere ideas are in fact sorry comforts in an unmanageable situation, they can be the beginning of demands, projects, even utopias, that enable people to organise in new ways to pursue them. The idea of human rights has gained much of its force this way, as a prism through which many efforts are focused and/

or refracted" (Jedediah Purdy, "Should we be suspicious of the anthropocene?," *Aeon Magazine*, March 31, 2015, http://aeon.co/magazine/science/should-we-be-suspicious-of-the-anthropocene/). We would argue that theories of animal rights can achieve the same force.

22. Jerry Anderson, "Protection for the Powerless: Political Economy History Lessons for the Animal Welfare Movement," *Stanford Journal of Animal Law and Policy* 4 (2011): 1–63.

23. Anderson, "Protection," 25. To cite a few statistics, 49,000 children aged 5–9 years were employed, 36% of boys aged 10–14 were working, of whom 76,000 were in the worst conditions of mines and factories; 20% of girls aged 10–14 were working, of whom 74,000 were employed in factories. Standard work days were 12–16 hours, in appalling conditions (ibid., 19).

24. And, in recent decades, the struggle for children's rights has moved beyond demands for protection and provision to demands for participation and full citizenship. We explore the relevance of these more recent dimensions of children's rights struggles for AR in Sue Donaldson and Will Kymlicka, "Rethinking membership and participation in an inclusive democracy: cognitive disability, children, animals," in *Disability and Political Theory*, ed. Barbara Arneil and Nancy Hirschmann (Cambridge: Cambridge University Press, forthcoming). Our focus in this chapter is on the lessons we can learn from earlier stages of the children's rights movement.

25. Anderson, "Protection," 6.

26. Phelps, *Changing the Game*, 8%. For a similar claim, see Melanie Joy, *Strategic Action for Animals* (New York: Lantern Books, 2008), 18. Neither considers the child labor movement.

27. The AR movement can learn from the history of the child labor movement about the importance of emphasizing nonexportation from the get-go. For example, consumer boycotts of animal flesh must occur in tandem with legislation to prevent export, or else the agriculture industry will simply shift to foreign markets, as we see in current efforts to create flesh markets in China and India. Similarly, bans on domestic practices (such as fur farms) must work in tandem with bans on imports. The child labor reform movement learned this strategic lesson the hard way, and has been playing catch up for a hundred years, although significant progress has also been achieved at the international level (Anderson, "Protection," 9).

28. Ibid., 8.

29. Ibid., 34.

30. Ibid., 38.

31. Ibid., 52.

32. Ibid., 55.

33. Norm Phelps, "Norm Phelps, Changing the Game" (blog), *Responsible Eating and Living*, July 30, 2013, http://responsibleeatingandliving.com/?page_id=11572.

34. We take the phrase from Blaire French, *The Ticking Tenure Clock* (Albany: SUNY Press, 1988).

35. Armory Starr, Luis Fernandez, and Christian Scholl, *Shutting Down the Streets: Political Violence and Social Control in the Global Era* (New York: NYU Press, 2011), 109. For other discussions of how the "Green Scare" (the targeting of AR and environmental groups as "eco-terrorists") has disrupted possible solidarities, see Peter Dauvergne and Genevieve Lebaron, *Protest Inc: The Corporatization of Activism* (London: Polity Press, 2014); Jeff Monaghan and Kevin Walby, "The Green Scare is Everywhere: The Importance of Cross-Movement Solidarity," *Upping the Anti* 6 (2008).

36. See Christiane Bailey, "Sexisme, Racisme et Spécisme: Intersections des oppressions" (paper presented at Philopolis, Montreal, February 7, 2015), available at: http://christianebailey.com/eventsevenements/sexisme-racisme-et-specisme-intersections-des-oppressions/.

37. Maneesha Deckha, "Disturbing images: PETA and the feminist ethics of animal advocacy," *Ethics & the Environment* 13/2 (2008): 35–76; Carol Glasser, "Tied oppressions: An analysis of how sexist imagery reinforces speciesist sentiment," *Brock Review* 12/1 (2011): 51–68.

38. As Dauvergne and Lebaron note in *Protest Inc*, this corporatization of activism is not at all unique to AR, and has multiple causes, not least state policies that have deliberately diminished the space for noncorporatized activism.

39. Sunaura Taylor, "Vegans, Freaks, and Animals: Toward a New Table Fellowship," *American Quarterly* 65/3 (2013): 757–764. See also Corey Wrenn's study of the images in vegan magazines, noting their preoccupation with images of young, slim, white women, and how this may exclude many groups in society ("An Analysis of Diversity in Nonhuman Animal Media", *Journal of Agricultural and Environmental Ethics,* 29/2 (2016): 143–65).

40. Licia Carlson, "Philosophers of Intellectual Disability: A Taxonomy," *Metaphilosophy* 40/3–4 (2009): 552.

41. Donaldson and Kymlicka, *Zoopolis*, chap. 2.

42. Carlson, "Philosophers"; Eva Feder Kittay, "At the Margins of Moral Personhood," *Ethics* 116/1 (2005): 100–131, and "Ideal Theory Bioethics and the Exclusion of People with Severe Cognitive Disabilities," in *Naturalized bioethics: Toward responsible knowing and practice*, ed. Hilde Lindemann, Marian Verkerk and Margaret Walker (Cambridge: Cambridge University Press, 2009).

43. That AR theorists would even use the term "marginal cases" is testament to a failure of solidarity with the disability movement.
44. pattrice jones describes this shared ground as follows:

> "there's a word for the idea that basic rights are dependent on particular capabilities: 'Ableism'. It's not merely that the most common rationale used to justify human dominion over animals is *similar* to the rationale used to excuse the mistreatment of people with disability. It is the *exact same* argument! *We have capabilities that they don't have, so we may rightfully force them to labor for little or no compensation. We have capabilities that they don't have, so we may rightfully restrict or control their reproduction. We have capabilities that they don't have, so we may rightfully confine them in institutions of total control. We have capabilities that they don't have, so we may rightfully kill them"*…. Rationality or reason is probably most frequently asserted to be a capability that both is exclusive to humans and somehow authorizes us to do whatever we like to other animals….Meanwhile, allegations of lack of rationality or the inability to reason have historically been used and continue to be used to deprive people of the most basic rights, including physical liberty and reproductive freedom." (pattrice jones, *The Oxen at the Intersection: A Collision* [Herndon, VA: Lantern Books, 2014], 68%)

> For a related view of the links between disability and animal rights, and their common agenda in challenging inherited ideas of dependency, the "natural," and the "normal," see Sunaura Taylor, "Beasts of Burden: Disability Studies and Animal Rights," *Qui Parle: Critical Humanities and Social Sciences* 19/2 (2011): 191–222.

45. This is a very condensed discussion of a complicated issue. For more extended discussion, see Will Kymlicka and Sue Donaldson, "Animal Rights, Multiculturalism and the Left," *Journal of Social Philosophy* 45/1 (2014): 116–135; and "Animal Rights and Aboriginal Rights", in Peter Sankoff, Vaughan Black and Katie Sykes (eds), *Canadian Perspectives on Animals and the Law* (Toronto: Irwin, 2015), 159-86. Claire Jean Kim, *Dangerous Crossings: Race, Species, and Nature in a Multicultural Age* (Cambridge: Cambridge University Press, 2015). A distinct but related issue concerns the strategic consequences of using "the dreaded comparison" between animal use and the Holocaust, slavery, and genocide. The danger, here again, is the instrumentalization of other groups' struggles. While such comparisons can and have been used in careful ways that help enrich our understanding of both human and animal oppression, Jews, blacks, and indigenous peoples understandably resent it when people appropriate their history in a purely instrumental way. As Kim Socha notes about PETA's use of comparisons with the enslavement of Africans and the

genocide of American Indians, "It is painfully obvious to any observer that PETA is not interested" in the experiences or struggles of people indigenous to the Americas and Africa: "they are just interested in the analogical possibilities of their subjugation. When white activists hijack other oppressions, they may face (not necessarily unwarranted) accusations of insensitivity" (Kim Socha, "The 'Dreaded Comparisons' and Speciesism: Leveling the Hierarchy of Suffering," in *Confronting Animal Exploitation: Grassroots Essays on Liberation and Veganism*, ed. Kim Socha and Sarahjane Blum [Jefferson, NC: McFarland & Company, 2013], 232). See also Angela Harris, "Should People of Color support animal rights?" *Animal Law Journal* 5/15 (2010): 15–32.

46. As pattrice jones notes, "Since racism is a key thread in the network of intersecting oppressions that include speciesism, and since sufficient expertise in the dynamics of race is a prerequisite for effective activism in a multicultural environment, such self-education should be seen not as something that takes times away from animal advocacy but an essential element of animal advocacy" ("Afterword: Flower Power," in Socha and Blum, *Confronting Animal Exploitation*, 273).

47. For discussions of animal resistance, see Jason Hribal, *Fear of the Animal Planet: The Hidden History of Animal Resistance* (Oakland: CounterPunch, 2010); and Dinesh Wadiwel, "Do Fish Resist?", *Cultural Studies Review* 22/1 (2016): 196–242. http://sydney.academia.edu/DineshWadiwel.

48. The main proponent of this abolitionist/extinctionist approach is Gary Francione. See, for example, "Animal Rights and Domesticated Nonhumans," *Animal Rights: The Abolitionist Approach* (2007), http://www.abolitionistapproach.com/animal-rights-and-domesticated-nonhumans/.

49. Donaldson and Kymlicka, *Zoopolis*, 77–89.

50. For another approach linking human sex/gender rights and animal rights, see pattrice jones, "Eros and the Mechanisms of Eco-Defense," in *Ecofeminism: Feminist Intersections with Other Animals & the Earth*, ed. Carol J Adams and Lori Gruen (New York: Bloomsbury, 2014), 91–108.

51. The promotion of this view is most closely associated with Gary Francione, who says: "We must be clear that veganism is the unequivocal baseline of anything that deserves to be called an 'animal rights' movement" ("The Paradigm Shift Requires Clarity about the Moral Baseline: Veganism," *Animal Rights: The Abolitionist Approach* [2012], http://www.abolitionistapproach.com/the-paradigm-shift-requires-clarity-about-the-moral-baseline-veganism).

52. Ian Smith, "Is There No Room for Rod Coronado in the Animal Rights Movement? The Problem with Veganism as the Moral Baseline," *Uncivilized Animals*, March 18, 2014, https://uncivilizedanimals.wordpress.com/2014/03/18/is-there-no-room-for-rod-coronado-in-

the-animal-rights-movement-the-problem-with-veganism-as-the-moral-baseline/. See also Adamas, "A Critique of Consumption-Centered Veganism," H.E.A.L.T.H. Web site, June 3, 2011, http://eco-health.blogspot.it/2011/06/crtique-of-consumption-centered.html; Jon Hochschartner, "Should a Meat-Eater Advocate for a Vegan Society?," and "What can animal activists learn from the free produce movement?," *Species and Class*, October 25, 2014, http://speciesandclass.com/2014/10/25/what-can-animal-activists-learn-from-the-free-produce-movement/.

53. http://collectivelyfree.org/team-page/

54. As pattrice jones notes, "the emphasis on personal veganism—and on persuading others to 'go vegan' as the sine qua non of animal advocacy—has created a chilly climate for people who might be ready to devote substantial time and energy to zoos, circuses, vivisection, fur or other forms of animal abuse and exploitation but are not yet ready to go vegan" ("Afterword: Flower Power," 266).

55. Anderson, "Protection"; Hochschartner, "What Can Animal Activists."

56. Anderson, "Protection."

57. Irene Blanken, Niels van de Ven, and Marcel Zeelenberg, "A Meta-Analytic Review of Moral Licensing," *Personality and Social Psychology Bulletin* 41/4 (2015): 540–558.

58. Phelps, *Changing the Game*. Kris Forkasiewicz argues that the preoccupation with personal veganism is problematic, not because it replicates a reactionary politics of personal virtue, but because it is devoid of politics entirely. As he puts it, "vegans thus join dumpster divers, urban gardeners, potluck organizers, 'lifestyle' anarchists, eco-villagers, and others who seem to turn their backs on the depressing political context of their milieu … The isolation is not a strategic or tactical choice; it is a surrender, an easy way out of a pressing situation, and a symptom of a serious crisis whose dynamic begins to make sense with a simple realization: the vegan movement sees animal exploitation not as a political matter, but as one of cultural and ethical misperception" ("Fragments of an Animalist Politics: Veganism and Liberation," in *Critical Animal Studies: Thinking the Unthinkable*, ed. John Sorenson [Toronto: Canadian Scholars Press, 2014], 49).

59. On how defining veganism as AR's moral baseline undermines alliances with indigenous peoples, see Dylan Powell, "Veganism in the Occupied Territories: Anti-Colonialism and Animal Liberation," March 1, 2014, http://dylanx-powell.com/2014/03/01/veganism-in-the-occupied-territories-anti-colonialism-and-animal-liberation/.

60. This is a consistent refrain in Kim Stallwood, *Growl: Life Lessons, Hard Truths, and Bold Strategies from an Animal Advocate* (New York: Lantern Books, 2014). Indeed, a surprising number of books on animal advocacy never once mention political parties.

61. The Party for the Animals (PvdD) in the Netherlands has had a significant impact on Dutch debate and policy, but this depended, in part, on its ability to get members elected into the national legislature (and now also into the EU Parliament) under Netherlands' unusually low threshold for proportional representation. In other countries, the option of a dedicated animal rights party may be less effective. On the PvdD's impact, see Simon Otjes, "Animal Party Politics in Parliament," in *Political Animals and Animal Politics*, ed. Marcel Wissenburg and David Schlosberg (London: Palgrave, 2014).

62. This fundamental point is too often neglected, even within AR theories, in part, because a surprising number of animal advocates seem to think that animals' ways of life are predetermined by their genetics, and hence they are not capable of leading different kinds of lives. We have elsewhere argued that domesticated animals are capable of many different types of social relationships with members of their own or other species, and of engaging in many different forms of cooperative activities including various forms of work. It is therefore a fundamental requirement of justice to create the conditions for animals to explore what kinds of lives they want to lead, and thereby exercise what we call "macro agency" (Donaldson and Kymlicka, "Rethinking Membership"). We explore how farm sanctuaries could provide a site for such macro agency in "Farmed Animal Sanctuaries: The Heart of the Movement? A Socio-Political Perspective," *Politics and Animals* 1 (2015): 50–74.

63. This engagement might extend to involving domesticated animals in the struggle for their liberation. Darren Chang explores this provocative idea in a recent talk about "Empowering 'Farm' Animals to Become Activists." https://www.youtube.com/watch?v=kg4LKc5LgrE.

64. DeFilippis, *Unmaking Goliath*; Hoover, "Human Right to Housing."

65. Justin Van Kleek, "The Sanctuary in Your Backyard: A New Model for Rescuing Farmed Animals," *Our Hen House*, June 24, 2014, http://www.ourhenhouse.org/2014/06/the-sanctuary-in-your-backyard-a-new-model-for-rescuing-farmed-animals/.

66. DeFilippis, *Unmaking Goliath*, discusses the Burlington Vermont community land trust model.

67. On denizen rights for liminal animals, see Donaldson and Kymlicka, *Zoopolis*, chap. 7.

68. And in the agricultural industry, animals are brutally exploited for reproductive services.

69. For example, Eden Alternative elder-centered communities aim to create "a Human Habitat where life revolves around close and continuing contact with plants, animals and children. It is these relationships that provide the young and old alike with a pathway to a life worth living" (Eden Alternative

Overview Brochure, 2014, available at: http://www.edenalt.org/word-press/wp-content/uploads/2014/02/Eden_Overview_092613LR.pdf). Intentional communities for people with disabilities provide another example of an integrated village approach to the development of supportive community, as an alternative to isolated homes or total institutions. See M. Randell and S. Cumella, "People with an intellectual disability living in an intentional community," *Journal of Intellectual Disability Research* 53/8 (2009): 716–726.

70. We have not touched on the health benefits of a plant-based diet, and how community activism around the right to healthy food provides another alliance opportunity for AR activists in cooperation with nutritionists and public health advocates.

71. The Harmony community in Florida represents an attempt, not without problems, to construct an intentional community that combines ecological and animal-friendly principles (Mona Seymour and Jennifer Wolch, "Toward zoöpolis? Innovation and contradiction in a conservation community," *Journal of Urbanism: International Research on Placemaking and Urban Sustainability* 2/3 [2009]: 215–236). On a smaller scale, many designers are figuring out ways to design cities to better accommodate liminal residents. See, for example, "Butterfly Bridge" and other projects by artist Natalie Jeremijenko, at http://www.nataliejeremijenko.com/about/.

72. Anderson, "Protection."

73. See Matthew Cole and Kate Stewart, *Our Children and Other Animals: The Cultural Construction of Human-Animal Relations in Childhood* (Farnham: Ashgate, 2014).

74. For a discussion of prospects and challenges of municipality-based legal advocacy for animals in the Canadian context, see Cameron Jefferies and Eran Kaplinsky, "Municipal Governance and Innovative Shark Conservation Efforts: Problems and Prospects," in *Canadian Perspectives on Animals and the Law*, ed. P. Sankoff, V. Black, and K. Sykes (Toronto: Irwin, 2015).

75. Donaldson and Kymlicka, "Rethinking Membership."

76. Laura Janara, "Human-Animal Governance and University Practice in Canada: A Problematizing Redescription," *Canadian Journal of Political Science*, 48/3 (2015): 647–73. See also Dan Lyons, *The Politics of Animal Experimentation* (London: Palgrave, 2013).

77. "Interview with John Sanbonmatsu" by Saryta Rodriguez, *Direct Action Everywhere*, December 16, 2014, http://directactioneverywhere.com/theliberationist/2014/12/1/interview-with-john-sanbonmatsu-associate-professor-of-philosophy-at-worcester-polytechnic-institute.

The Problem of Akrasia: Moral Cultivation and Socio-Political Resistance

Elisa Aaltola

Socrates famously asserted that immorality is ignorance, and that with an adequate amount of knowledge one will always do what is good. The rationalism inherent in Western philosophy and culture has internalised this notion, with at least two practical implications: first, it is presumed that one's moral choices are rational (rather than emotive, culturally produced, or purely ambiguous); second, it is thought that in order to effect a moral/political change in others, all one has to do is provide rational arguments and information.

In the context of debates concerning the status of nonhuman animals, these implications are evident. Arguably, most individuals tend to assume that their choices related to other animals are rational; hence, meat-eaters suppose that meat-eating can be rationally defended, while overlooking the manner in which the consumption of animal products may be grounded on emotive, cultural, or utterly ambiguous justification. Animal advocates, on the contrary, often presume that in order to advocate veganism, all that is required is more rationally produced information concerning the treatment and moral value of nonhuman animals. Thus, both those who

E. Aaltola (✉)
Department of Social Sciences, University of Eastern Finland, Kuopio Campus,
Kuopio 70211, Finland
e-mail: elanaa@utu.fi

© The Author(s) 2016
P. Cavalieri (ed.), *Philosophy and the Politics of Animal Liberation*,
DOI 10.1057/978-1-137-52120-0_5

117

consume other animals and those who seek an end to such consumption tend to believe in the significance and even centrality of reason.

Yet, there are grounds to argue that such belief in rationalism is unwarranted. First, studies in social psychology suggest that the majority of moral judgements are based on intuitions, that is, immediate knowledge, which, in turn, tends to be grounded on either emotions or cultural stereotypes. Indeed, rationality may primarily serve an "epiphenomenal role" in human morality, inasmuch as we tend to use it after judgement, in order to justify our intuitions post hoc.[1] This means that the meat-eater who is fully certain of the rationality of her choices may actually be following emotive reactions towards animals (for instance, contempt), which, in turn, may be sparked or accentuated by cultural stereotypes ("pigs are dirty and dumb"), and that she only seeks to offer rationales for those choices after they have taken place. What emerges with this picture, then, is the perplexing possibility of a wholly nonrational stance on our treatment of other animals—a phenomenon that has, in the recent years, been termed as the "meat-eater's paradox". Within the realm of this paradox, Western people both love and eat animals more than ever before. Concern for animals is rising alongside animal consumption and industrial farming, and the reasons stem from collective, culturally produced psychological mechanisms, rather than rationality.[2] Thus, although individuals tend to presume that their choices concerning other animals are grounded on reason, in practice they may be quite nonrational.

Second, if this applies, animal advocacy may be placing over-emphasis on reason. Although rationally synthesised information and arguments are pivotal for nonbiased moral understanding (the type of understanding that sees value also in those others who are not similar, close, or useful to us), they do not suffice, and reflection on the role of emotionally and culturally produced intuitions is also required, if one is to effect a change amongst those who eat pigs and love dogs. That is, if rationality plays a relatively limited part in practical, de facto moral judgement, offering ever more (and only) rationally produced reasons for change is futile. Arguably, this is sensed by many activists, who are frustrated over the slowness of change, and over the stubbornness of those who choose not to alter their ways of relating to the nonhuman world. The activist may feel that although all the information and arguments concerning the moral status and treatment of other animals are there, society carries on its routine instrumentalisation of animal others as if it had never seen the information and never heard the arguments, and the activist is pained by this behaviour. She may ask: How can a widespread change ever become possible?

Therefore, a tension between rationality and nonrationality lingers beneath the surface of the debate concerning animal consumption. The chapter explores this tension via a paradox that has puzzled Western philosophy since its birth: akrasia. Akrasia is a state in which one rationally recognises that x is wrong, but still does x. Hence, one can rationally recognise the validity of the animal ethics arguments, and rationally think one ought to become vegan, but still persist in meat-eating, milk-drinking, and egg ingestion as if that rational recognition had not fully penetrated one's consciousness. What will here be termed "omnivore's akrasia" differs from "meat-eater's paradox" in that it concentrates on the conflict between reason and action, rather than between emotion and action; moreover, it explores the phenomenon via philosophy, rather than psychology.

The aim of this contribution, in the context of the debate fostered by the present volume is to reconcile the increasing awareness of non- or less-rational tendencies within us with the view that rational, moral reflection is still required in our attitude towards nonhuman animals. The chapter attempts to achieve this by exploring both the type of emotions and desires, which may cause one to ignore rational reflection, and the possibility of cultivating that reflection by paying attention to the wider forces that support akrasia. Particularly the latter—the socio-political forces that impact and often hinder our capacity for reasoned moral reflection—will be given emphasis, as it will be suggested that human beings become non-rational in their moral choices, when the surrounding society does not support reflective forms of ethics.

RATIONALISTS AND THE DEMAND FOR CULTIVATION

In order to explore omnivore's akrasia, the rather evident first port of call is rationalism, the philosophical school, which has, perhaps, had the most profound impact on both the Western culture and psyche, and the dualism between humans and other animals. For rationalists, human beings are essentially and primarily rational creatures, and it is this specific inclination that is presumed to set culture apart from nonhuman nature, and humans apart from nonhuman animals. For rationalists, it is pivotal that one's relation towards not only the world in general, but also more specifically nonhuman animals, be grounded on reason; hence, omnivore's akrasia rises as a bewildering, embarrassing phenomenon (how can human beings, who are supposedly "above" other animals due to their reasoning abilities, nonetheless, forsake those abilities in their treatment of animals?). Yet, classic rationalists from past eras have their explanations for how akratic oddities may take place, and these explanations hold promise of illuminating novel directions via which to explore contemporary akrasia in the context of animal politics.[3]

Plato discussed akrasia in some length in both *Protagoras* and *Gorgias*, and asserted that although we will always seek to do what is "good", our grasp of that good can become obscured due to false appearances that mask "evil" as "goodness". These appearances, again, are spurred by nonrational desires and misleading appetites (wants) and passions (emotions), the proto-example being violence, which, in a jealous, needy rage, can gain the appearance of a moral necessity. Hence, Plato agrees with Socrates in claiming that as long as we grasp the "good" with clarity, we will follow its lead; however, he also noted that such clarity is easy to lose.[4] If we want something, and if we have emotions that accentuate that want, we may become overwhelmed by a desire for the evil, whereby our mind masks it as good. In the context of omnivore's akrasia, this would suggest that individuals who recognise the validity of the arguments for veganism, and who see the pieces of footage from industrial farms, may, nonetheless, keep consuming milk, eggs, and flesh, because they both want those things and harbour negative emotions (contempt, superiority, disgust) towards other animals, and thereby form such a strong desire for animal products that animal consumption gains the appearance of goodness. Here, a state of mental muddle overtakes reason, as one may still note the validity of animal ethics arguments, and yet perceive meat-eating, wholly irrationally, as good. In other words, clarity is lost, and baffling layers of wants and emotions veil violence as moral, as one continues to, in Plato's words, "wander all over the place in confusion".[5]

Plato's remedy is clear: self-control and cultivation. One is to control appetites and passions, and thereby to avoid situations, in which nonrational desires can spark misleading appearances. One is to strive towards becoming good, to pay attention to what types of wants and emotions one follows. The solution for omnivore's akrasia, then, would be to reflect on one's own being—what we are like as creatures, what motivates us. This virtue ethics approach requires attention to *oneself*, rather than merely on rational information or arguments concerning *others*—we are to pause, to observe, and to scrutinise our own wants and emotions, our own tendencies to follow mere appearances rather than the truth. Following

Plato's famous cave allegory, the aim is thereby to cease concentrating on shadows on the walls of the cave, and to turn one's attention towards the light that causes those shadows—as painful as this can be. Here, the "shadows" are the appearances offered to us by the surrounding culture: the animal imagery put forward by the media, and the imagery concerning animal consumption continually constructed and repeated by machineries of marketing that wish for us to purchase ever more pieces of nonhuman bodies. The "light", on the other hand, is the capacity for reflection, and ultimately the "goodness" which that reflection will manifest.

Thereby, omnivore's akrasia could be overcome by emphasising the need for self-control and cultivation. In practice, this would mean the decisive will to reflect on what type of creatures animals are, and what their moral status demands, regardless of the emotions and wants within us, often pushed forward by the surrounding culture. Yet, such a solution is demanding. Notions of self-control and cultivation may appear uncomfortable, negative, and even ridiculous to contemporary individuals. This is because current, particularly Western, societies are rested on forces that resist any accentuation of those notions, and, indeed, stand as antithetical to them. The consumerist ethos, so prominent within contemporary social realms, is that of short-sighted, hedonistic egoism, within which we are pushed to become ever more impulsive, desiring, wanting, and emotive, so that marketing can easily mould us into creatures of purchase. In short, we become beings, in whom the cacophony of advertisement can cause emotions and wants, and who impulsively follow those emotions and wants, unwilling to pause, to reflect, and to consider one's aims and motives, let alone the justification for one's actions. Arguably, the sort of cultivation suggested by Plato is thereby alien to the contemporary world, in which consumerism actively resists refinement towards pausing, reflective moral thought. Hence, it is also alien to the context of animal consumption, particularly when we note that it is precisely *consumption*, the consumerist tenet, which tends to define the politics revolving around the treatment of other animals. What Plato is asking is thus quite challenging to contemporary consumers of meat, eggs, and milk.

Still, the notions of cultivation and control stand as all the more pivotal. As Alasdair MacIntyre famously suggested, what the current era—and indeed much of modern philosophy—has forgotten is the core of virtue ethics, without which human animality cannot thrive: the question what we *could* be, if we were to follow our potential. Much emphasis is placed on what we are, and what we should do, but the key factor between these issues—and indeed one that glues them together—is missing.[6] Human telos, potential, is left without attention, and as a result we all too eagerly merely hang on to that what is, and insist that we are to be accepted as we are, forgiven all despite our actions, loved in all our faults without any notion of change—and that we are to be allowed to follow whatever wants, emotions, and desires we may happen to have. It is, arguably, within this ethos that omnivore's akrasia also flourishes—an ethos, which insists that we are to "accept" the suffering of the world. "So it is", the milk-drinker may say melancholically, with certain sadness towards nonhuman plight, while ingesting a liquid that causes that very plight. The task, then, is to bring cultivation back, to render the effort to make oneself more reflectively moral, something positive rather than dull, and a way of manifesting human potential rather than a hindrance.

One source for rethinking the possibilities of cultivation comes from another rationalist, René Descartes, the philosopher usually (and rightly) perceived, due to his dualism, in less than positive light within the realm of animal ethics, who also discussed akrasia. Like Plato, Descartes argued that "akratic breaks" are instigated by misleading emotions and desires. He also underlined, among other things, the role played by mental limitations, which prevent one from perceiving morality with clarity, and the power of societal habits and customs in turning our attention away from that which is rationally good. As for Plato, for Descartes too the solution was cultivation. For him, it meant "suspension of the will": when something is causing us to get confused and akratic, we are to wait for clarity and regroup our moral reflection.[7] Here, Descartes shares some striking similarities with contemporary discussion, as psychology and psychiatry have tried to map out various cognitive shortcomings or limitations that render akrasia into a reality, and as they have accentuated the relevance of succumbing to public opinion (habit, custom) as one source of akrasia. Hence, it has been argued that shortcomings/limitations such as impulsivity, apathy, and a state of ambivalence (within which we cannot decide between different forms of information) enforce akrasia by rendering one incapable of controlling one's wants and emotions; moreover, the suggestion is that within a state of ambivalence, people tend to follow the opinion of the majority (and thereby habit).[8]

Therefore, many act against rationally produced moral arguments not only because of wants and emotions but also because of cognitive tendencies and societal habits; indeed, perhaps, wants and emotions entwine with these further catalysts, whereby, for instance, impulsivity and apathy may coincide with given emotions that justify their own continued existence (for instance, contempt may feed moral apathy). Importantly, from the viewpoint of cultivation, limitations and habits may entwine. That is, perhaps societal customs and habits can so shape our psyches that we become particular types of individuals. Hence, the consumerist society may render us ever more impulsive with its constant sway of marketing mechanisms that evoke us to buy and act without thinking, and the information society entangled with it may mould us into more cognitively ambivalent, indecisive creatures with its continuous, overwhelming cacophony of often contradictory and fragmented pieces of "information". In the latter case, the reality becomes utterly ambiguous, and one falls into a state of anxious uncertainty (even when some pieces of "information", some arguments, are more rational than others). In both cases, we become more morally and politically apathetic, too lethargic to take actions that would help others. Ultimately, it may be societal habits, the society's discursive infrastructures, which position akrasia all too often as our very core. This, again, implies that in order to eradicate akrasia, we need to scrutinise social habit, or, more broadly, the various forms of societal, political infrastructures, which affect our behaviour.

Irrespective of his views on the moral status of nonhuman animals, Descartes' stance is relevant also to omnivore's akrasia. Perhaps the akratic milk-drinker is not only persuaded by wants and emotions to follow non-rational moralities. Perhaps she is also sculpted by the surrounding societal and political habits and institutions into a creature who wants to buy dairy and meat even when she rationally knows she ought to refrain, who remains indecisive even when handed an animal ethics argument she rationally recognises as valid, and who struggles with apathy. Perhaps, then, societal foundations (marketing as their spearhead) are rendering human individuals evermore cognitively prone towards omnivore's akrasia.

Therefore, what we can learn from the rationalists is the following. First, self-control and cultivation are required—two notions often viewed with utter suspicion in the contemporary culture eager to assert that we ought to concentrate on our present desires rather than on the distant consequences of our actions, and that talk of "reason" is preachy and reference to "self-control" harsh and merciless. Eroding akrasia requires that one seeks to become more orientated towards the rationally reflected "good", that one cultivates oneself away from impulsive, unreflected emotions and desires, and that, thereby, one becomes less apathetic in pursuing the good. Here, the former akratic meat-eater would wilfully do what is moral by keeping impulses, emotions, wants, desires, and senses of ambivalence and apathy in check; she would be motivated by the good, and eager to materialise it in her own actions. Second, such a solution may appear overly idealistic, for there are more powers at stake than our own will to do what is right. These powers come from the societal level, which is often antithetical to efforts of cultivation, and eager to shape us as ever more impulsive creatures, filled with egoistic, hedonistic desires and accompanying emotions, and ultimately defined by moral and political ambivalence and apathy—also (and, perhaps, particularly) in our relations towards nonhuman animals.

If cultivation is required, so, then, is wider societal change away from those infrastructures that are currently reinforcing the akratic slumber. This leads us away from the rationalists, onto socio-political takes on akrasia.

THE SOCIETAL POLITICS BEHIND AKRASIA

Amelie Rorty has investigated akrasia from the viewpoint of its societal context. Invested in the Aristotelian tradition, within which human individuals gain their political and moral understanding—and ultimately their very humanity—only in the company of other such individuals and wider societal ramifications, she has posited that akrasia can only be understood in reference to societal, political milieus. Thereby, akrasia felt at the level of individual psychology "reflects conflicts among the larger social and economic institutions".[9] Indeed, akrasia cannot be individualised, rendered into a phenomenon stemming purely from our own cognition, but finds support—and at times is wholly instigated by—the social and political dimensions surrounding us. Interestingly, the Aristotelian interpretation of akrasia also reminds us of the role that societal institutions have on enabling (rather than hindering, as is often currently the case) moral cultivation; that is, the social, political modes of being around us ought to support our path towards virtue, our efforts to fulfil our potential and telos, and thereby our struggles to become more reflectively moral.

Rorty is careful to insist that for Aristotle, morality takes place within broader social, political ramifications (*nomoi*)—being moral requires constant interaction with other moral agents. The impetus for this claim is found precisely from the cultivating aspect of such interaction: we become capable of cultivating ourselves towards evermore virtuous modes of being, if we get feedback from others, and if we are given platforms on which to reflect on and discuss what is good. As Rorty posits, Aristotle is referring not only to direct interaction with other individuals, but also to the manner in which societal infrastructures bear an effect on cultivation. For him (and for her), various social factors, from legislation and economics to education, at best facilitate the refinement of the citizens' character traits, and, indeed, gain one substantial justification for their own existence in this cultivating force.[10] The society appears as a realm, which supports the aim portrayed within virtue ethics as the core of human life: to become more virtuous and moral, to become more good.

Hence, the Aristotelian emphasis on society and its integral ties with our moral agency, our virtue, offers a way to approach cultivation from a broader perspective than afforded by the rationalistic tradition. Although many rationalists—not only the mentioned Plato and Descartes, but also, for instance, Spinoza—saw links between doxa (public opinion) and akrasia, they still tended to reduce the latter into individual human psyches. This is because, although they grasped the connections between the *birth* of akrasia and wider societal spheres, they tended to leave the *solution* of akrasia in the hands of the individual—her capacity for self-control and to reorientate towards reason with the sheer power of her will. The Aristotelian framework reminds us that individuals do not exist in solitude, and thereby may not be able to solve akrasia merely via use of their own power. Rather, they are partly born out of societal configurations, and require the support of those configurations in order to combat akratic states.

Therefore, omnivore's akrasia may require more than individual effort. It appears evident that simply stating "you ought to have self-control!" does usually not suffice in sparking moral change. Reminders of self-control are important, even pivotal in this era of passive hedonism, but they require support from our wider societal settings and institutions. Simply put, as long as nonhuman animals are categorised by different societal institutions and spheres as fleshy, milky, tasty pieces of commodity, it is difficult (even if clearly not impossible) for individual human beings to follow the arguments for animal liberation. It is here that akrasia becomes most tangible: one knows the arguments, but one is enticed to buy that ice-cream, purchase that packet of bacon, because societal institutions, including education and health, economy, and, particularly, marketing, either offer us irrational views as rational and valid, or tell us that regardless of what we know, we ought to just go ahead and ignore the ethically obvious (obvious, if we spend any time reflecting on it) in the name of financial and culinary gain, or customary habit. In order to affect moral and political change towards veganism in individuals, attention thereby needs to be directed onto their social contexts, with particular emphasis on the following question: How can society facilitate moral cultivation in relation to nonhuman animals?

This requires three things: (1) we no longer view akrasia as limited to individual omnivore psyches, (2) we note the societal mechanisms/habits, which cause akrasia, and (3) we pay attention to how resistance and change may be sparked.

Internalism Versus Externalism

Therefore, Rorty's Aristotelian stance has three implications. First, akrasia cannot be solved merely by limiting attention to individual moral psychology—it requires positioning that psychology within societal, political ramifications. Second, current moral theory (arguably, including that found in animal ethics) is often ignorant or dismissive of the relevance of wider societal and political landscapes in enabling moral cultivation. Third, whereas contemporary social structures tend to stand as antithetical to cultivation, alternative structures, capable of supporting cultivation and eradicating akratic apathy, need to be mapped out.

Rorty is keen to underline the first of these points. She criticises a type of psychological individualism still evident in contemporary sciences and culture, which suggests that our decisions (including moral decisions) are grounded on the inner worlds found within our individuality—our personal emotions, desires, and so forth. As suggested earlier, according to Rorty, moral psychological aspects of individual life cannot be reduced to the individual, but rather find origins from the surrounding socio-political arrangements: indeed, the latter constitute the most profound basis of character formation. Simply put, and following externalism our individual psychologies partly stem from societal ramifications and from political spheres, and hence moral decisions need to be evaluated against not only the individual but also the societal. As Rorty suggests, "Individual akrasia may indicate social disorder"... "Akrasia is not always a solitary activity: it often works through sustaining social support."[11] Therefore, akrasia becomes a social issue, one to be addressed by paying attention to societal and political structures (such as legal, financial, or educational institutions), rather than by merely focusing on our inner worlds and our inner cognitive or emotive tendencies. The cure for akrasia consists, thereby, of reconfiguring social institutions in a direction, where they support cultivation towards virtuous forms of life, the "good".

Hence, Rorty offers support for the claim made earlier in relation to Descartes: that our wider cultural, social, and political contexts impact the constitution of our psyches, and thereby strengthen akrasia. Just as our beliefs are often structured and reinforced by social realities, so too are personality dispositions, which feed akrasia. It is on these grounds that akrasia emerges as a social, political phenomenon—it is not to be reduced to individual personas, to personal psychological processes, but rather takes flight from *nomoi*—and equally, it can only be cured via shifting attention from the personal onto the public.

Diego Romaioli et al. have also brought forward the societal/political ramifications of akrasia, and criticised the tendency to reduce akrasia into individual, internal psyches (a tendency from hereon called "internalism"). They suggest that akrasia may, in fact, be a pseudo-problem, in that it does not exist in the form usually assumed in philosophical literature. Whereas the latter presupposes akrasia to be confined to the individual subject, it may be that akrasia takes place within broader contexts external to the subject.[12] Like Rorty, they claim that the key, underlying mistake is to presume that moral judgements are wholly internal.

Romaioli et al. warn us against an approach dominant in Western philosophy, within which mental states are universalised, codified, materialised, and reduced into psychiatric algorithms in order to render them technically understandable. This, advances the above internalism, according to which we are closed units of mentation, atomistic individuals separated from our surroundings, inside which psychological phenomena take place, largely (or even wholly) unaffected by broader contexts.[13] Combined, these tenets suggest that we are subjects whose mindedness takes place within (rather than in relation to others or the broader external ramifications), and whose mental states can be perceived as stemming from something akin to machineries or programmes. For Romaioli et al., this modernist depiction of mindedness is flawed on a foundational level. Our mind and its constitution are not enclosed from contact with others; instead, mental states occur in relation to those others, often emerging as something inherently motivated and sparked by the external (the social, the political), as we continually negotiate intentions, judgements, and so forth in relation to what lies beyond our "selves". Thus, Romaioli et al. push forward an image of the "mind understood as a combination of nodes within an interpersonal network that is in state of perpetual change, depending on the context and objectives".[14] Our mental states, including moral judgements, are context-dependent, as they change and alter on the grounds of the identity of those who surround us. Indeed, not only our mental states and judgements but also our identities and the very sense of self fluctuate depending on the social context, as we are constantly reworking who we are, and hence manifesting a multiplicity of identities: "Despite giving the subjective and mistaken impression of being consistent and integrated, each human being's identity is complicated and disconnected, in fact more like a society of the mind."[15]

This, again, is significant from the viewpoint of akrasia, because as we are not internal, algorithmic programmes of reason, our judgement often deviates from the rational, and even wilfully ignores it. Instead of aiming at rationality, it can vary drastically, depending on the social context—who we are engaging with, what we are surrounded by, and so forth. Romaioli et al. argue that, indeed, in each episode of social interaction, a new sense of reality emerges, a specific world of meaning, which sparks differences in judgement. Since "knowledge is never self-contained, but always comes about as the result of intense dialogue with other consciousnesses",[16] akrasia does not arise from internal, private conflicts or contradictions, but rather emerges as a pseudo-problem precisely because it may not involve a contradiction as such, but merely reflects the heavily contextual nature of human mentation.

Hence, Romaioli et al. end up supporting Rorty's claim: instead of consisting of internal judgements, our moral psychologies are formed in relation to the external and the social, and thereby seeking to solve contradictions like akrasia requires awareness of this social dimension. Now, Romaioli et al. make a mistake in forsaking cultivation: after noting how socially embedded our minds and akrasia are, they appear to follow relativism, and assume that we ought to simply accept this embeddedness without further scrutiny (and thereby accept akrasia)—it is precisely on the grounds of forsaking reason that they term akrasia a "pseudo-problem". What remains missing is a grasp of the importance of our rational side (no matter how frail it may be), and the need to cultivate that side towards more morally reflective directions. Hence, instead of the relativism endorsed by Romaioli et al, it is Rorty's Aristotelian aim of cultivating virtue (including rational virtue) that ought to be emphasised. Yet, despite these failings, Romaioli et al. offer important insights into both the cultural presumptions surrounding internalism, and its inevitable failure—insights worthy of attention also in the context of nonhuman animals.

Indeed, all the aforementioned help in making sense of omnivore's akrasia. First, such akrasia is often construed in internalist terms, and thus people may view any societal efforts to cultivate moral reflection towards animals with hostility, as manifestations of malign paternalism. Here, internalism combines with individualism, as it is suggested that everyone ought to decide their consumptive habits independently, individually, without being given gentle nudges from wider societal settings (one manifestation of this is the hostility towards "meatless Mondays" evident in many countries). Omnivore's akrasia may also be entwined with the presumption that one is acting rationally even when one knows this not to be the case—the image of human beings as algorithmic centres of reason is, perhaps, so strong that even when they wilfully ignore reasoned arguments, many still feel justified, as if their actions *must* (somewhat mysteriously) have a rational basis. Arguably, this is an inherent part of omnivore's akrasia—the feeling that eating lamb or fish must, on some level, be wholly rational, since human beings have been doing it for millennia. Yet, despite these presuppositions of internalistic rationality, the akrates is flowing with societal settings, continually altering her perceptions in the contextual vein suggested by Romaioli et al. Her takes on other animals gain different perspectives, accentuations, and nuances, depending on the context (where and to whom she is speaking; whether she is invested in a macho work environment, or simply flowing within the identities pushed forward by the society of consumption). Hence, it may appear as self-evident, as utterly clear and "natural", that one ought to eat other animals despite whatever logical arguments against doing so one may be offered. This forms the core of omnivore's akrasia, the strong sense felt by the akrates that she is to follow various, contextually altering anthropocentric rationales, even when they are not rational.

Secondly, and more importantly, the tenets put forward by Rorty and Romaioli are evident in the mentality of animal ethics and advocacy. Arguably, there has been a strong tendency within these fields to presume that minds are internally constructed, and that all that is required in order to affect a change is to have an impact on those internal landscapes. Hence, it is often supposed that human individuals are ontologically atomistic, self-contained units of mentation, who are clearly separated from each other and their surroundings, and who are perfectly capable of creating and altering their own moral judgements when offered sufficient information. In the meantime, the role of the social in instigating forms of moral stances towards nonhuman animals tends to either be ignored, or brushed under the broad concept "anthropocentrism". What animal ethicists and advocates tend to do, then, is to offer logical arguments or points of information, or (less commonly) push forward moral emotions, instead of noting the role of the societal/political dimensions.[17] Moreover, it is presumed that minds, indeed, are something akin to algorithms, which one may alter with suitable programmes (logics) or other factors (emotions)— at the extreme of this tendency stands the new sort of animal advocacy, which uses the techniques of marketing in order to spark changes in our supposed algorithms, as if we truly were operable creatures, who can be successfully manipulated as long as our inner mechanisms are known.[18]

What surfaces, then, is an advocate's view of the meat-eater and the milk-drinker as an internally confused being, whose moral choices are (at least predominantly) individually produced, and whose (algorithmic) confusion can be easily cleared via offering him or her logical, informative tools. The social and the political rarely enter the scene.

Hence, much of animal ethics and advocacy is based on a particular, internalistic conception of moral psychology—one that is rested on atomism and algorithms. What all too easily remains without notice is the manner in which our moral stances towards nonhuman animals are largely shaped by our surroundings, namely, cultural meanings, language, the media, the education system, the legal system, the financial system, consumerism, marketing, and so forth. This has an impact on the efficacy of animal ethics and advocacy. There are formidable social institutions and compelling networks of cultural meaning, which assert that nonhuman animals are things to be used, and they will keep igniting and sustaining akrasia in many with ease, when their only rivals are reasoned arguments and information. Indeed, animal ethics and activism are fighting not only colossal forces, but also forces with succinct, optimising insights into the human psyche, some of which—particularly consumerism and marketing—deliberately and directly seek to increase consumption of nonhuman animals. Thus, there are socio-political instances that actively, intentionally shape omnivore's akrasia, and without paying attention to these instances, and conversely, to the potential role of the social in supporting cultivation, animal ethics and advocacy will remain frustrated.

Now, none of the aforementioned implies that rational arguments are to be discarded.[19] They constitute, along with emotions such as empathy, our moral sense in its ideal form; without rational, logical analyses and emotive dimensions entwined with these analyses, we would fall into mindless cultural relativism, and in so doing would lose our potential for morality and virtue. What this suggests, however, is that rationality (or emotion) alone does not suffice. In order to affect widespread political change in how nonhuman animals are treated, valued, and related to, attention needs to be placed *also* on the societal, and particularly on how the societal may be cultivated in a manner that enables our own moral cultivation—for even rationality and emotions do not exist purely internally, but rather are affected by and reverberated in the social.[20]

Rationality Versus Societal Habit

Rorty also repeats a theme brought forward by Descartes (albeit without reference to him), as she discusses the relation between akrasia and habit. After taking pains to accentuate the fact that akrasia may not always be based on the self, but may equally be arbitrary, stemming from forces outside the individual, Rorty considers the role of societal habit as the source of akrasia. This is significant, because, as Rorty emphasises, habit tends to be nonrational. Whereas the standard stance on human mentation presumes mindedness to be primarily grounded on the complexities of reason, it may, instead, find much of its substance from nonrational habit. Hence, akrasia may neither be directly based on any conscious beliefs nor be justified via such beliefs: it simply takes place, without being rested on propositional expressions. From this perspective, it is intuitive, rather than rationally, propositionally defendable.[21]

Here, Rorty takes a step away from Plato and his tendency to underline the akratic role of desires. As seen, within Plato's description, akrasia appears rational, the muddling surfaces of false appearances mask it as "good", and this is because the desire behind it pushes us to deem it as a reasoned way to act. Arguably, akrasia often consists of egoistic desires, giving our actions the false appearance of rationality (and indeed, those actions can be rational in the optimising sense, which refers to using reason as a tool to satisfy wants). Yet, such is not always the case. Although rational optimising, fuelled by egoism, appears as one obvious motivation for akrasia (one simply chooses to pursue hedonism in order to serve the most immediate needs of the self, instead of following arguments that would require revising those needs), our akratic choices may take place beyond the possibility of conscious deliberation, and thereby beyond the possibility of conscious hedonism or egoism; they may originate from collective, habitual ramifications, which serve no clear purpose for any one individual. Importantly, this unconsciousness concerning the causes of our actions and choices may be so extensive that we cannot offer any reasoned expression for our akratic states.[22] When habit invites and pushes us to ignore reasoned moral arguments, it does not necessarily replace those arguments with its own logic, but simply creates an argumentative void, within which we keep going against the good without any capacity to justify ourselves. As Rorty notes, this void appears to be related to the loss of history that defines many collective, cultural habits: we no longer know the original rationale or motivation for those habits (for instance, survival) and carry on repeating them nonetheless. Habits lose their original meaning, and become empty, often mindless routine, which defines our moral choices.

Here, Rorty is offering an important reminder of the potential mindlessness of akrasia. Akrasia is not necessarily constituted by optimisation based on serving egoistic, hedonistic desires, wants, or emotions. Instead, it may be wholly lacking any argumentative support, and wholly separate from egoistic, hedonistic gain. Thus, it is nonrational both in origin and in justification. Here, we begin to act in ways, the motivations and contents of which we remain oblivious to. Habits formed by others, perhaps a long time (even millennia) ago, are internalised and pursued, even when they include no rationale, and serve no clear purpose. Rorty argues that these habits have an astoundingly strong hold of us, to a point where we are willing to forsake our individually reflected beliefs in order to flow with publically produced opinion. Hence, many participate in collective acts, which they individually would never condone.[23]

Like Rorty, Romaioli et al. suggest that the background assumption of the depiction of internal mentation, offered in the previous section, is ego-centrism. Within it, the ego is confined to itself, and its mental states are directed solely by it; moreover, the ego is presumed to direct those mental states with machine-like arrangements, acting in response to rational judgements that optimise self-benefit, motivated by its own, internal, self-directed aims (if not overtaken by malfunctions such as sudden emotive disruptions). However, since minds do not consist of internal, logical mechanisms, this ego-centrism is flawed.[24] What Rorty and Romaioli posit is that egoistic rationalisation is often presumed rather than factual: that it is a specific way of explaining human behaviour and akrasia, which may not always hold explanatory power, and that akrasia may also be wholly mindless and nonegoistic.

Again, relevance to omnivore's akrasia is clear. It is precisely the nonreflected, habitual presumptions and ways of relating that may play a pivotal role in determining omnivore's akrasia. Instead of viewing that akrasia as solely consisting of short-sighted egoism, which prioritises the hedonistic emotions and wants of individual human beings, often sparked by surrounding societal institutions, it is beneficial to note the nonconscious, nonoptimising, even nonexpressible as cause for one's judgements and actions. This is not to say that egoism plays no part, or that the consumerist society refrains from tapping into and creating hedonistic, self-directed desires, wants, and emotions in us; as will be suggested shortly, consumerism and marketing continuously evoke self-regarding hedonism in human beings, to a point of, perhaps, making many ever more narcissistic, and thereby ever more prone towards ignoring the needs of nonhuman animals. However, it is equally misleading to reduce everything to such ego-accentuating mechanics. Often, omnivore's akrasia may be grounded on intuitions, which again originate from societal habit—a habitual way of evaluating, defining, and treating nonhuman animals in a given way. These intuitions tend to be nonexplicable, and hence the akrates will struggle to justify herself (for her, it "just is" self-evidently so, that other animals can be used for food or entertainment). They may also remain wholly nonoptimising, nonrational, and fail to serve any egoistic benefit—thus, the akrates will carry on consuming milk, eggs, and dairy even when it is unhealthy for her, is ruining her environment, and is causing discomfort and guilt within. Indeed, it can be argued that veganism serves human benefit the best, and that many, are becoming increasingly aware of why this should be so—and yet, akrasia persists. The background reason may be habit, which has long ago lost its rationale (survival) and has become simply a nonrational, even destructive mode of action, empty of its original content.

In order to combat akrasia, attention needs to be directed, thereby, not only on the ego-serving desires, wants, and emotions supported or produced by societal, particularly consumerist, institutions, but also on the nonconscious habits, which push people to view other animals as commodities even when, individually, those same people may accept the arguments offered by animal ethics and be vehemently against causing nonhuman animals suffering. Arguably, many fall in this category: they consider particularly industrial farming immoral, they would never kill an animal for food with their own hands, and they are not particularly hedonistic, eager to shout out "steak is good!" as a defence for meat-eating;

yet, they, nevertheless, are lulled into the collective habit of animal consumption, thus continually, day after day, acting against their own moral reason. It is here that we meet the nonhedonistic, nonegoistic akrates, who feels guilty over eating hamburgers, but who, nonetheless, even helplessly, feels impelled to do so.

Hence, perhaps, omnivore's akrasia is partly constructed on loss of history, within which reasons for judgement and action are buried under the repetition of habit, and individuals are swayed under floods of collective behaviour, incapable of asserting their own responsibility or even moral agency. The loss of history signals, thereby, a loss of reason, a state of passivity, and a type of mindlessness, within which individuals follow collective habits, even when those habits go against their own rationality, their own reasoned ethics. It is via the mindlessness of habit that collective violence gains much of its momentum.

Yet, as hinted earlier, one should not wholly sideline the factuality of rationalising egoism. Indeed, even when it is part of a cultural narrative, and is thus culturally produced (the modernistic notion of the self-sufficient, rationally optimising human being), it often materialises in us, becoming evoked by those narratives and acts of construction. Here, what was an element within a false and mythical internalist description of mentation (egoistic, algorithmic calculation) becomes, via the repetition of that description, at least a partial reality as we live the myths we invent. Hence, akrasia can be interrelated with rationalising egoism: we nonreflectively pursue our own gratification, and follow whatever modes of behaviour enable us to rationally optimise our behaviour so as to achieve that gratification. Significantly, reason is used, but not reflectively; it concerns the best ways to achieve our own gain, not the justification or nature of aiming for that gain. Human beings appear as creatures, who can easily use their wit to serve ego-centric ends, but who are wholly unwilling to use that wit in order to reflect on what those ends are, what they should be, and what forms of gaining them are morally sound. Mindlessness may, thus, walk hand in hand with ego-orientated optimisation.

Such egoistic optimisation can, in fact, coincide and intertwine not only with mindlessness but also with habit—indeed, it adds to the lure of the latter. Since we are not urged to reflect on our ends and ways of achieving those ends, it is easiest to follow preset modes of behaviour, in the belief that the society and its customs know best which way we ought to go. In short, many give authority to societal habit, presuming it to offer the most rational, soundest way of serving one's ends. Thus, echoing the existentialist notion of "bad faith" (within which we surrender our freedom to choose, our capacity to use our own reflection, to the hubbub of social custom that tells us what we ought to do, what we ought to be like, and within which we thereby lose our sense of "authenticity"—our capacity to make responsible choices—and become "alienated" from others and ourselves), this perspective on akrasia suggests that instead of reflective, multi-layered rationality, it is ego-centric optimising which motivates us—and that this ego-centric optimising is thought to be best secured by social habit, which we mindlessly, as if willing to shift authority away from ourselves, follow. Via ego-centricism, we thereby paradoxically risk yielding power over our own choices, our own moral judgements, to mindless habit.

Indeed, the consumerist aspects and institutions of the society impel individuals ever more strongly to hand over authority to societal habit. Consumerism reveals the connections between egoism and the willingness to let go of reflective (rather than purely optimising) rationality: reflective reason would force one to view critically the messages of marketing, the contents of entertainment-orientated media, and to thereby ultimately scrutinise critically the basis of one's very egoism. In short, it would stand as antithetical to the mindless urge to pursue instant gratifications, and as such would take authority away from cultural habit, from the consumerist ethos itself. If consumerism is rendering us ever more mindless, ever more habit-like, ever more alienated, reflective rationality has the potential to save us. Yet, since consumerism has infiltrated most of our institutions, and since its survival is dependent on lack of reflective rationality, it appears evermore unlikely that reflection will flourish. Hence, perhaps, consumerism is enforcing in many ever more egoistic, and thereby ever more prone to ignore reflection in relation to our choices and judgements; that is, consumerism is enforcing in many a mindless eagerness to follow habit, a state of continual akrasia.

Again, these stand as central in the analysis of omnivore's akrasia. Consumerism forms the most evident societal reason for the increase in animal consumption, and within this process, both egoism and habits may play a pivotal part. Here, it is not egoism in any essential, inherent sense, which leads to omnivore's akrasia, but rather culturally, societally produced egoism, which stems from the image of the mind highlighted by modernity, and which is actualised with particular force within consumerist realms. The human being becomes a unit of rational optimisation rather than reflection, and simultaneously she begins to follow societal habit in the belief that it secures that optimisation, that it is based on a clear rationale. Hence, the meat-eater will declare that despite all animal ethics arguments, all creatures have the right to concentrate on themselves, that looking after one's own benefit is the way of the world, even if the costs of doing so are a regrettable cause for sadness. She will also let marketing and the media tell her, how that benefit is realised, and will thereby partake in endless habits of animal consumption in the belief that they *must be* rationally beneficial, perhaps even reflectively good. It is often for such reasons that a person familiar with the catastrophic costs and consequences of meat-eating still tucks into steaks, believing that "it must be rational and good" if the surrounding society depicts it so. Here, the akratic omnivore has given authority to societal habit, and has become reflectively mindless, alienated from his or her own choices.

Therefore, whereas many would like to believe their moral choices to be rational, they tend to be grounded on habit; similarly, akrasia may stem from habit rather than hedonistic, egoistic optimisation. However, the existence of the latter needs to be noted, for although some akratic meat-eaters carry on their consumptive habits without egoism, others position egoism as their motive for following those habits. In both cases, rationality as reflection, as moral deliberation, is missing, and hence omnivore's akrasia becomes a form of mindlessness, a state of bad faith. The societal enters the individual via eradicating his or her desire for reflective reason, and via introducing habit at its place.

Yet, it must be noted that there are dangers to depicting a passive image of human beings as creatures, who can easily be pushed towards whatever directions social institutions wish them to go. This is because such a depiction implies hopelessness, and it is precisely hope, which is required for social, political change. But, how are such hope and change—and thereby the eradication of omnivore's akrasia—achievable?

Contradiction and Resistance

Rorty's aim is to direct attention to the failures inherent to those of our societal/political instances, which support akrasia. In an Aristotelian vein, she calls for a critical reevaluation of societal and political institutions, as it comes to their involvement in akrasia.[25] Indeed, this stands as the key concern within the socio-political approach towards akrasia: how to alter those institutions, which cause or uphold akratic modes of behaviour?

Rorty highlights contradiction as one key failure, which ought to be paid attention to. Within these contradictions, societal/political institutions simultaneously both encourage and condemn the same actions. We are told to pursue x, and not to pursue x, and here "x" can stand for a wide variety of values common within contemporary societies. Rorty clarifies, "While promoting habits of co-operation, they [social institutions and economic systems] also reward radical independence; while condemning aggression, they also praise 'aggressive initiative'; while admiring selfless devotion, they also reward canny self-interest."[26] In short, we are encouraged, on the surface, to follow given virtues, whilst concurrently being pushed towards acts that are antithetical to those virtues; in the liberal society, one is to believe in universal equality whilst buying clothes from chains that use slave labour, and one is to believe in solidarity whilst participating in economic processes that feed elitism. This contradictory nature is inherent in contemporary economical processes and other institutions, and it allocates the same action differently, depending on the sphere of the society in which it takes place. The contradiction can be transparent (an individual person stealing is "greedy", a corporation stealing is "successful"), but it can also be so deeply rooted as to become almost untraceable—and in any case, it is rarely discussed or reflected on.[27]

Following Rorty's logic, what surfaces as the core tension in most of our institutions concerns, arguably, moral ideology. On the one hand, societal/political institutions advocate other-directedness, the notion that we are to pay heed to others, to fight for equality, to offer assistance and support, to be even self-sacrificing and altruistic; on the other hand, those same institutions demand that we concentrate on ourselves, accept and intensify inequality, and concentrate on our own rights rather than duties towards others. Hence, we are told to be other-directed and utterly self-directed in one and the same sentence. The other-directed ideology exists predominantly in the private realm of self-identity and personal relations. Our social sense of self is constructed on the idea that we are good, other-oriented, moral creatures who believe in solidarity, equality, and self-sacrificing, even heroic concern. The self-directed ideology of (neo) liberalism, on the contrary, exists primarily in the public realm of work, finance, and consumption, and manifests itself in the presumption that one is to work hard in order to be able to consume hard, and that one is to concentrate on one's own immediate wants (whatever those wants may be) rather than the more distant needs of others. Here, we almost unwittingly, without noticing, adopt an identity based on greed, individualism, egoism, and hedonism, and with it a way of approaching others, which tends to remain blind to the needs and value of those others.

Indeed, Rorty suggests that it is particularly the self-directed approach, which is proving influential, as various institutions push their egoistic tendencies into our personal psychologies. For her, the most powerful of these is the economy, as "our motives and habits are profoundly influenced by the way that economics drives civic politics. Both, taken together, pervade absolutely every nook and cranny of our lives."[28] Here, the human psyche begins to resemble the economy; it internalises the presumptions of egoism, inequalities, and reward inherent to market relations. The moral ideological split that exists within the social realm is, perhaps, thereby increasingly being decided in the favour of the ego-directed mode of perceiving the world and its creatures (with implications also from the viewpoint of the previous section).

This contradiction must stay hidden, however. In order not to reveal it, conformation to unspoken rules is necessary: we are meant to intuit in which contexts given type of behaviours are acceptable, and when faced with conflicts, remain silent with regard to the ensuing, uncomfortable sense of incongruity, should one emerge.[29] In short, one is to follow set norms that change according to context, ignore contradictions in one's behaviour and identity, and thereby accept conflict and remain uncritical, unreflective. Hence, it "just is" so that other-directedness can be celebrated by those who simultaneously remain utterly egoistic. It is here that akrasia comes in. We begin to act in ways that we would not condone on the level of reasoned judgement, and thus we participate in hedonistic behaviours, for instance, whilst remaining resolutely opposed to hedonism. Thereby, we internalise contradiction and conflict, a mode of being within which we rationally advocate freedom, and practically support slavery.

Romaioli et al. also speak of such contradictions and efforts to conceal them. They suggest that there are rituals, with which individuals are driven to reconcile different, conflicting identities so as not to become aware of the possibility of contradictions. The aim of those rituals is the blurring of the lines between different identities—differences are quite simply obscured. Yet, the sudden obviousness of conflicts is a constantly looming possibility, and when they do become obvious, the typical course of action is to resort to narratives, with which to absolve us from responsibility, and with which to maintain a sense of continuum and harmony in the face of the conflict; indeed, this is how akrasia is most often dealt with. If all else fails, we espouse that we simply were not "ourselves" when the supposed akrasia took place; here, intentions are separated from actions, and the self is affiliated with the former instead of the latter.[30]

These contradictions stand as the key to reconfiguring those institutions that are involved in akrasia. It is by exposing the conflicts, making them apparent, that akrasia and its root causes can be brought to the fore, and ultimately eradicated. Thereby, if we wish to erode akrasia by mapping out its socio-political sources, attention needs to be placed on the contradictions which those sources contain. The contradictions need to be made visible, and their societal, political roots an object of critical scrutiny—it is here that resistance against akrasia becomes possible.

Now, resistance rarely occurs, for the need to conform is potent, and the notion of breaking the rules by noting contradictions raises shame and pacifies. Rorty offers a damning report: "Workplaces, banks, courts, armies, and hospitals all stream, direct, and constrain citizen motives. They define flow charts of duties and virtues, rights and obligations whose infractions carry severe costs and sanctions. Nonconformists are regarded with suspicion; they are charged with irrationality; they have difficulty eliciting cooperation and they suffer from pressure that is intended to produce guilt, or at the very least, shame."[31] As a consequence, there is a tendency to view surrounding social and political forces as authoritative, and thereby unquestionable. Yet, resistance is necessary. As Rorty posits, the akrates cannot recover from the contradictory state without noting the contradiction, and ultimately without political reform (particularly including a renewed stance on market relations): "If Daemona [the essay's main character] were to recognize the extent to which her desires have been manipulated, realizing exactly how socioeconomic policies violate her more general aims, she might be better positioned to check her akrasia."[32] In conclusion, Rorty suggests that overcoming akrasia requires both personal reflection and, most of all, social and societal support—starting from rendering contradictions evident and igniting political reform.

The cognitive dissonance between having moral regard for and eating animals is something continually supported by societal institutions: welfare legislations and the media tell us how important it is to care, and marketing tells us how important it is to eat (those we care about). Indeed, often the same social institution offers two completely contradictory beliefs, and thus the media can both bring forth a cavalcade of cute animals, who are suggested to deserve nothing but moral concern (e.g. think of "pet TV-shows"), whilst also sprinkling those cavalcades with notions of animality, which posit that other animals are sheer instrumental material to be consumed by humans (think of meat or dairy adverts, in the middle of the aforementioned shows). Following a similar contradictory logic, advertisement can both tells us how important it is to care via using "welfare" as a selling point, and simultaneously accentuate how animals are raw material designed for human consumption, and hence worthy of little or no moral concern. These are the dissonances and contradictions, which lead to the meat-eater's paradox, —and they also pave the way for omnivore's akrasia. Within the latter, one thereby recognises the validity of animal ethics arguments and advocacy information, but nonetheless, wholly incongruously, carries on consuming pigs and chicken, as if no contradiction, in fact, was taking place. Many have, thereby, internalised the ideological split between other-directedness on the one hand, and self-directedness on the other; societally produced contradictions become assimilated into personal psyches, and they are no longer noted for what they are.

The contradictions are grounded on different sets of context. First, there is the set of contexts based on *use-value*. The existence, the ontology of nonhuman animals is classified according to categories that are structured on the basis of their use, and these categories tell us how to relate to those animals. It is the category, the context, rather than the animal herself, which guides our ways of relating to her; it is here that utter contradictions begin to accrue, as one may view a rabbit wholly differently, depending on whether the rabbit is perceived to be "a companion animal", "a laboratory animal", "a farmed animal", "a vermin", or "a prey". The moral status of other animals is made to follow the contours of cultural categories, and these categories tend to be utterly contradictory, resulting in constant tension and conflict; yet, within the akratic logic, we are not meant to notice these contradictions. Second, there is the set of contexts based on *moral value*. Here, it is various moral values, which are used to illuminate the treatment of other animals, again often in a wholly contradictory sense. Thus, we are meant to "care", to have moral concern, and to speak of "inherent value", "respect", and "justice" in relation to nonhuman animals, but equally we are meant to follow the anthropocentric norms, which erase, or at least marginalise, those notions and replace them with hierarchies, exclusions, utilisation, egoism, and instrumental value. These differing moral perspectives tend to be completely contradictory, but we are meant to move between them without noting those contradictions, and, thus, one can watch Babe on television whilst digging into a bacon sandwich. Third, there is the set of contexts based on *identity*. Here, our take on nonhuman animals can alter radically, depending on whom we are speaking to, whether we are at work, in the public sphere, in our own home, in a macho environment, and so forth. One may express love for a companion dog at one's home, and support vivisection on dogs in the public sphere, as a representative of the education system; equally, one may emphasise care towards animals among friends, and advocate brutal callousness among peers defined by masculinity.

These sets of contexts (and there surely are more) are continually utilised by different socio-political institutions, ranging from the legal system to education, party politics, and economy. In relation to animal welfare legislations, the legal system is grounded on utter contradictions in how nonhuman animals are treated; in one context, poisoning a dog to death or keeping it in a tiny cage is legal, whilst in another context, such behaviour would result in penalisation. Within education, children are taught to care, to feel even awe at the face of nonhuman animals, but they are simultaneously taught to view those same animals as self-evident objects of consumption. Within various levels of the economy and market relations, contradictions are particularly obvious, as the commodification of nonhuman animals has entered most aspects of both human and nonhuman lives (even love is commodified, as all aspects of taking care of companion animals are made into sellable objects, and as even respectful encounters with wildlife increasingly come with price tags). Party politics echoes the economy, and hence politicians are eager to pledge for "welfare" whilst condoning the ever-increasing intensification of animal use; again, the different discourses produced in party politics reverberate in the public sphere, and ultimately often in the private consciousness. The message is completely contradictory: have moral concern, be good towards nonhuman animals, ignore moral concern, and view those animals as sheer instruments for human use.

Yet, as suggested earlier, the contradictions are silenced, as one is meant to navigate between the different ways of relating to nonhuman animals without noticing their incongruous nature. We are guided to intuitively know, how to act in different contexts, and to smoothly shift between those contexts, even when our ways of acting are completely contradictory. Thereby, criticism is rarely offered, as it becomes simply, quietly accepted that moral concern is for companion animals, and slaughter houses for "production" animals. Indeed, these contradictions are arguably internalised, as they come to signify animality for us: the living, breathing, sentient creatures that nonhuman animals are (the specific cows, pigs, and fishes of this world), are replaced with cultural categories, which again define both them and how they are related to. Since the categories are contradictory, our sense of animality becomes inherently contradictory, as if this was inevitable and even natural. When contradictions and conflicts do emerge, and when the uncomfortable recognition that something is amiss is, thus, potentially sparked into existence, various narratives are used to tone them down, thereby preventing their full recognition. When all else fails, within these narratives, blame is allocated elsewhere ("politicians/industries/consumers should ensure animals are treated well"), with ultimately nobody emerging as willing to accept responsibility.

These forms of contradiction and cover-up enable omnivore's akrasia. When surrounded by a social world, within which akratic decisions in relation to nonhuman beings are the norm rather than the exception, it is astoundingly easy to be lulled into states, within which one goes against what one rationally knows to be justified. Resistance, the willingness to expose these contradictions and to critically evaluate them, is desperately needed, but all too rare. Vegans and animal advocates are easily shunned as nonconformists, labelled as "preachy" or even "terrorist". They gather suspicion; they are told they ought to feel guilt and shame for going against the order of things (and some assimilate that shame, feeling embarrassed, socially conscious, and awkward, when raising the issue). Within such a scenario, it is not surprising that many choose not to become vegan for the simple reason that they feel doing so would marginalise them (at the work place, amongst associates and peers, even within their own family, etc.). Resistance is, thus, met with hostility, and the hostility again ensures that many keep their akratic ways of life, knowing they ought not to consume other animals, but still unable to materialise that knowledge into actions in the fear of becoming a social "oddity", or even an "outcast".

Hence, more resistance is required. For omnivore's akrasia to become a less prominent frustration for animal ethics and activists, and most importantly for it to cease being the source of suffering and death for billions of nonhuman animals, the contradictions evident within the socio-political institutions need to be brought to the fore. Moreover, these institutions need to be altered and reshaped in ways that erase contradiction. Perhaps, human life will always be contextual, confusing, and contradictory; perhaps the societal cannot (nor should) be structured with nothing but logics—but when it comes to the well-being and life of other cognitive, sentient individuals, such contradictions require reevaluation, and our societal structures a dose of rational, moral reflection.

CONCLUSION: TOWARDS CULTIVATION

For the rationalists, akrasia arises from irrational emotions, wants, and desires (Plato), together with cognitive defects and social habits (Descartes). The way to eradicate akrasia is cultivation, so that we become more capable of using our rational, reflective thought when making moral decisions. From this viewpoint, omnivore's akrasia emerges as a state, within which one is overtaken, for instance, by contempt towards other animals, desire for given, learned culinary choices, impulsivity or moral apathy, or even simply learned custom, within which consumption of other animals appears as unquestionable, even when it goes against moral reason. The aim also for the meat-eater or the milk-drinker is to practice self-control, to cultivate oneself so as to become more capable of reflection, and less vulnerable to the pressures posed by emotions, wants, desires, defects, and habits.

However, whereas the rationalists tend to approach akrasia as a failure of the individual psyche, a more societally aware stance on akrasia requires that its mechanisms be also sought from those wider societal spheres, within which individuals make their moral choices. As explicated by Romaioli et al. and Rorty, minds never make decisions purely internally, in a vacuum; rather, akrasia stems partly from the socio-political institutions, which surround us. Thereby, if we wish to become more reflective, if we wish to advocate cultivation so as to eradicate omnivore's akrasia, attention needs to be placed on the social processes that affect our thoughts and emotions, intuitions and actions. The question becomes as follows: how to push social institutions towards a direction that better enables and cultivates moral reflection and use of reason, whilst diminishing akrasia?

Three elements appear important. First, recognition that our moral choices are not wholly internal; second, noting the way in which societally produced habit (egoistic or nonegoistic) often takes over reason, thereby rendering our choices nonrational; and third, awareness of the contradictions within societal/political institutions, which support akrasia. The solution to omnivore's akrasia is thereby to pay heed to the social layers that instigate akrasia, to the animal consuming habits that are continually pushed forward by different sections of the society, from education to consumerism and the economy, and to bring forward the contradictions inherent to these habits. In short, the task at hand is to render evident what has remained largely unexplored—the way in which societal institutions invite omnivore's akrasia via habits, which rest on contradictions, and which, thereby, invite simultaneously both moral recognition of, and utter moral disregard towards, nonhuman animals.

Ultimately, eradicating omnivore's akrasia requires that societal institutions from education to media and the economics are altered towards a way that invites moral reflection and is already structurally grounded on equality among embodied, sentient creatures. The aim ought to be the bringing forward of the human potential for moral reflection, the cultivation of our ability to be other-orientated, morally reflective creatures—a potential which cannot materialise without also taking into account other than human animals. Hence, we need Platonic pursuit of virtue and Cartesian "suspension of the will" when it comes to other animals. Social institutions must be evaluated and restructured from the viewpoint of how they support such cultivation, and how they thereby enable moral agency in relation to other beings. This, again, requires a return to the Aristotelian way of viewing societal, political entities as something that ought to continually support our moral growth (rather than, as is often the case currently, stunt our moral agency). In practice, such an undertaking means that, for instance (and particularly), the consumerist society, and its tendency to reduce trillions of nonhuman creatures into sheer, marketable commodities, ought to be critically viewed within animal ethics and advocacy.

This may portray an overly grim picture of the current state of affairs. Of course, more and more people are becoming vegan, and making such a choice is becoming increasingly easy. Yet, next to the growing invitingness of the vegan way of life, stand the forces antithetical to it, and amongst those forces astounding quantities of people, for whom the vegan choice appears impossible even when they note the validity of the relevant arguments. In order to render omnivore's akrasia less prominent, and in order to make it the rare exception rather than the excepted norm, attention needs to be placed on the societal structures that currently portray it as a feasible, and even as an enticing escape from moral demands. That is, animal ethics and advocacy must be rendered more societally, politically aware; as this collective volume shows, hopefully paving the way for further initiatives, they must move emphasis away from internalism towards recognition of how the external impacts our values related to other animals, and they must learn to more fully demand change within societal structures, to the point of being openly critical of given institutions. It is only then that moral cultivation on the individual level can become more widespread. Within the utopian society, not only individuals but also social institutions would be structured on moral reflection and the ideal of cultivation, the notion of becoming ever more virtuous and generous towards others, also when those others are nonhuman animals.

Finally, it ought to be noted that, perhaps, akrasia is an inevitable part of human life. We cannot always follow reason, for such perfectionism would place unwarranted demands on our everyday life, and distance us from other significant aspects of moral existence (such as more emotive, intuitive ways of approaching others—ways, which, of course, define much of what we are, are integral even to the use of reason, and are never to be marginalised, even if reflection on their nature and their justification is continually required). However, there are situations in which reason is elemental, and in which akrasia cannot be condoned. These are situations, which have to do with the lives and well-being of other sentient, morally significant beings—not only humans, but pigs and cows, pikes and herrings, seagulls and lizard, rats and bees.

NOTES

1. Jonathan Haidt, "The emotional dog and its rational tail: A social intuitionist approach to moral judgment", *Psychological Review* 108 (2001): 814–834.
2. Melanie Joy, *Why We Love Dogs, Eat Pigs, and Wear Cows* (New York: Red Wheel/Weiser, 2009).
3. I have written more extensively on rationalists and akrasia in "Empathy, and Psychology of Animal Protection Movements", *Politics and Animals* 1 (2015).
4. C.W. Taylor, "Plato, Hare and Davidson on Akrasia", *Mind* 89 (1980): 499–518; Thomas Brickhouse and Nicholas Smith, "Socrates on Akrasia, Knowledge, and the Power of Appearance", in *Akrasia in Greek Philosophy*, ed. Christopher Bobonich and Pierre Destree (Leiden: Brill, 2007).
5. *Protagoras* 356d6–356d7.
6. Alasdair MacIntyre, *After Virtue: A Study in Moral Theory* (Indiana: University of Notre Dame Press, 1984).
7. René Descartes, *On Method* and *Passions of the Soul*; Byron Williston, "Akrasia and the passions in Descartes", *British Journal for the History of Philosophy* 7 (1988): 33–55.
8. Annemarie Kalis, Andreas Mojzisch, Sophie Schweizer, and Stefan Kaiser, "Weakness of will, akrasia, and the neuropsychiatry of decision making: An interdisciplinary perspective", *Cognitive, Affective & Behavioural Neuroscience* 8 (2008): 402–417.
9. Amélie Rorty, "The Social and Political Sources of Akrasia", *Ethics* 107 (1997): 654.
10. Aristotle, *Nicomachean Ethics*; Rorty, "The Social and Political Sources of Akrasia".
11. Rorty, "The Social and Political Sources of Akrasia", 649, 651.
12. Diego Romaioli, Elena Faccio, and Alessandro Salvini, "On Acting Against One's Best Judgment: A Social Constructionist Interpretation for the Akrasia Problem", *Journal for the Theory of Social Behaviour* 38 (2008): 179–192.

13. A third, equally potent tenet states that there are universally understand-able and objectively definable "best judgements" which one ought to follow. Here, the subject scrutinises his or her beliefs, positions them in hierarchical orders, and pursues the one positioned at the top-end of the hierarchy. If the mind is seen to follow the order of mechanical natural sciences, judgement is seen to reflect the order of economics, wherein one logically and rationally weighs the potential benefits of each option before coming to a well-reflected, objectively defensible conclusion (Romaioli et al., "On Acting Against One's Best Judgment"). However, in this critique of reason, it seems that Romaioli et al. are throwing too much away simply in order to manifest that minds are also socially produced, that is, they are unnecessarily eroding the role of rationality.

14. Romaioli et al., "On Acting Against One's Best Judgment", 185.

15. Ibid., 186. They continue: "If we are deeply involved with the world at large, in relationships with others, in life's vicissitudes, why should we then continue to believe that anxiety and problematic behaviour originate from an individual's mind?" Ibid, 189.

16. Ibid, 187.

17. Of course, there are exceptions, among which are ecofeminists and the Care Tradition, which has sought to map out the gender politics behind human-nonhuman relations.

18. Here, internalism is mixed with the view that internal landscapes can be externally manipulated, if this is done consciously, knowing the mecha-nisms of those landscapes.

19. Although, of course, the aforementioned implies that the rationalistic approach to akrasia, *a la* Plato and Descartes, is also vulnerable to internal-ism and atomism. As long as solutions to akrasia are mapped out solely from reasoning ability, or other internal mental features, and if they hence remain exclusive of the recognition of the importance of the societal and political in taking part in the shaping of human minds, omnivore's akrasia will keep its hold on anthropocentric psyches.

20. Of course, there is nothing radical, per se, in noting that minds are partly shaped by their social environments—this is what humanities and social sciences have been often vocally highlighting in the contemporary era. The notion also has considerable philosophical support from schools such as phenomenology, which have underlined how our embodied mindedness (Maurice Merleau-Ponty's chiasmatic bodymind) refutes the possibility of purely internal mentation. As embodied beings, we exist in relation to our surroundings, and for such social, affective creatures, that relation often consists of intersubjective states with others, wherein the borders of where one's mind ends and another's begins can become difficult (and unneces-sary) to determine. Thus, for instance, Max Scheler poignantly criticised

the depiction of human beings as purely internal, reason-oriented creatures. Calling the view, according to which we are imprisoned into our minds, atomistic, he sought to explicate how we and our mentation are, instead, formed in relation to the world and its beings, and thus inherently open to the traces, which the social can leave in its wake (Max Scheler, *The nature of sympathy* [London: Routledge, 1997]).

Yet, the critique of internalism does, nonetheless, retain some radicalness in the context of even philosophy, within which there still exists a tendency to accentuate the human mind as a singular, enclosed entity, formed of reasoning ability and language. Indeed, this tendency towards a type of mystic glorification of the human mind as a highly developed rational, propositional machinery, is quite evident in most sciences—even when those sciences otherwise note the relevance of the external and the social. Thereby, although it is vigorously noted that the social alters the individual, it is simultaneously suggested—to a point of this still maintaining its place as, perhaps, the most commanding of cultural paradigms—that human beings are inherently rational, independently thinking, internally defined creatures. Therefore, a type of cognitive dissonance appears to stand at the root of even humanities and social sciences. On the one hand, humans are made into creatures largely defined by the social, and thereby vulnerable and passive, and on the other hand, they are depicted and even glorified as rational beings, capable of such cognitive, internal processes that they are termed the "pinnacles" of evolution, and hence ultimately viewed as utterly active creatures, at times seemingly nearing omnipotency, with their rational, scientific capacity.

21. Ibid.
22. Ibid.
23. Ibid.
24. Romaioli et al., "On Acting Against One's Best Judgment".
25. Rorty, "The Social and Political Sources of Akrasia".
26. Ibid., 652.
27. Ibid. Rorty uses the example of envy, which is officially condemned, but on which marketing and the mass media heavily relies—we are made to feel less, to form comparisons between ourselves and others, via advertisement and junk TV, so that we can be fed ever more wants, satisfiable only via a never-ending process of spending.
28. Ibid., 654.
29. Ibid.
30. Romaioli et al., "On Acting Against One's Best Judgment".
31. Rorty, "The Social and Political Sources of Akrasia", 654.
32. Ibid., 657.

Animal Subjects and the Logic of Human Domination

Brian Luke

INTRODUCTION

Nonhuman animals are the direct targets of slaughter, blood sport, vivisection, and ritual sacrifice. We, who seek the abolition of these and other such institutions, naturally turn to human liberation movements, some of which have achieved notable successes, for strategic and tactical inspiration. The very means by which we describe the movement (animal *rights*, animal *liberation*), its goals (*abolition* vs. *reform*, for instance), and the treatment being opposed (*oppression*, *exploitation*), deploying terminology characteristic of human protest movements, indicate how implanted is our assumption that the struggle for animals is fundamentally akin to previous intra-human campaigns, such as the abolition of slavery and the enfranchisement of women.

But the movements to abolish meat production, sport hunting, animal experimentation, and blood sacrifice are not fully analogous to movements against male dominance, white supremacy, militarism, imperialism,

B. Luke (✉)
First United Methodist Church,
207 South Court Street, Marysville,
OH, 43040, USA
e-mail: lukebria@gmail.com

© The Author(s) 2016
P. Cavalieri (ed.), *Philosophy and the Politics of Animal Liberation*,
DOI 10.1057/978-1-137-52120-0_6

or capitalism. Even as we avail ourselves of the full range of tactics used by these campaigns—direct action, civil disobedience, lobbying, referenda, petitions, lawsuits, boycotts, consciousness raising, and so on—we must acknowledge that this struggle is different, with differences, perhaps, significant enough to require novel strategies and tactics. Simply borrowing from human liberation movements, assuming that what worked there will work here, may not be sufficient.

Three differences are immediately apparent.

First, intra-human justice movements require leadership from the dominated group itself. Though it's a mistake to assume that nonhuman animals are never political, and that they cannot struggle against their own oppression, the image of ceding leadership of the animal liberation movement to nonhuman animals, in some analogy, say, to the realization that men directing the fight against patriarchy necessarily stunts and distorts feminism, appears more likely to retard progress for animals rather than to advance it. This first disanalogy with human protest movements warns us against any simple equation of viable strategy and tactics.

A second disanalogy refers to end goals. Even if there is a definable set of "human rights" that covers what all humans need to flourish (a dubious proposition), we certainly cannot assume that those human rights—life, liberty, and the pursuit of happiness, for instance—are just what we should be aiming for with regard to nonhumans. Respect for others requires acknowledgment and acceptance of their differences, and in the case of inter-species justice and morality, there are likely to be at least as many definitions of what counts as liberation as there are species themselves.[1] There is not one animal liberation movement, but many, as the kinds of life and their various forms of exploitation by humankind are many. The very range of what counts as "flourishing" as we consider nonhumans in all their diversity, from eukarya to bacteria to protists, chordates to mollusks to arthropods, mammals to birds to fish, cats to dogs to horses, alerts us that we may need a correspondingly wide range of strategies and tactics, surpassing those developed within human-to-human struggles.

The third disanalogy provides the topic of the present chapter. Humans and nonhumans do not occupy the same place within the logic of human domination. By "human domination" I refer simultaneously to systems within which one group of people dominates another, and systems within which people dominate nonhuman animals. It is useful to have an overarching term of this sort because the various systems of dominance are interwoven, at times mutually supportive, and can be seen to form an interlocking global system. It is not just that men dominate women, whites dominate people of color, owners dominate workers, and people dominate animals, each system operating in isolation from the others; rather, racism draws on sexism for support, and vice versa; capitalism makes use of sexism and racism to further its development; the systematic abuse of animals is furthered by feminizing them, while at the same time women and blacks are put down by considering them as nonhuman; and so on.

Each of the various systems of domination has its own history, its own peculiarities, its own conflicts and alliances with the other systems. The structure and development of human domination, like human history in general, being complicated, confusing, and constantly evolving, it becomes simplistic and ahistorical to claim "a" logic of human domination, as if there is just one, forever and for always.

But patterns do recur.

Such as, calling someone an animal when preparing to lock him up in a cage, as in the following news item:

> A Pennsylvania man involved in a long-running property dispute with local officials pleaded guilty on Friday in the fatal shooting of three people at a 2013 municipal meeting and was given three life terms in prison. Rockne Newell, 61, of Saylorsburg, also received an additional 61 to 122 years in prison, prosecutors said. E. David Christine Jr., the district attorney in Monroe County, called Newell an "animal" who belongs in prison. His lawyers said Newell has a history of mental illness.[2]

More broadly, entire groups can be deemed animalistic, as a means of legitimizing their systematic exploitation when it has come under challenge. Whether blunt or subtle, this "bestialization" is there, all the time, driving debates about intra-human justice, because the logic is there, underpinning legitimation of the status quo:

Premise 1.   Of course we exploit animals. It's natural.
Premise 2.   Those people over there ... well, they're like animals, aren't they?
Conclusion.   So of course we exploit those people. It's natural.

Within this syllogism, animals occupy a "privileged" position (one of their few privileges in Western society): their exploitation is the given, the part of the argument that does not need its own support. The naturalness of animal exploitation surrounds us all the time, part of the air we breathe.

One can find examples where this logic is flipped, where, for instance, the domination of women, workers, or children is taken for granted, and the abuse of animals is defended in terms of that. I'm not arguing that the logical priority of animal exploitation is eternal, transcending all context, but rather that within the context at issue—intra-human protest movements looked to as potential source of animal rights strategy and tactics—animals are negatively "privileged," that is, their exploitation by humans is understood as natural by all parties.

Both revolutionaries and reactionaries do it: reactionaries, as they seek to preserve the status quo by reinforcing the "animal" nature of the subordinate group; revolutionaries, as they work for liberation by striving to "humanize" their constituents. A fundamental strategy of human justice movements is to elevate the status of the subordinate group by portraying them as "fully human," that is, not like those beasts over there, whom, of course, we will continue exploiting because ... that's just what you do to beasts. This basic, recurring strategy of human justice movements is detrimental to the cause of animal liberation and, of course, unavailable for us to borrow due to its inherent logic.

In her call to animal liberationists for new strategies and tactics, Paola Cavalieri maintains that strategy must be "addressed to the basic logic of the system to be attacked."[3] In this chapter I discuss four prominent works within the Western literary tradition, focused on four different areas of human domination: the ritual sacrifice of children in ancient Greece, men's subjugation of wives in Elizabethan England, slavery in the antebellum south, and the exploitation of wage labor in early twentieth-century Chicago. In each case I consider how animals are positioned, logically and systemically, as the narrative depicts its particular intra-human institution of focus. I then conclude this chapter by drawing some general lessons for movement strategy.

FOUR CASE STUDIES IN THE LOGIC OF HUMAN DOMINATION

I. Iphigenia at Aulis

The myth of Iphigenia, recounted in Euripides' tragic drama *Iphigenia at Aulis* (408 BC) and other sources, is as follows. Greek forces are at harbor in Aulis, waiting for wind sufficient to sail across the sea to Troy, so they can sack the city in retaliation for the abduction of Helen, wife of Menelaus. A prophet informs the expedition's leader, Menelaus' brother

Agamemnon, that the wind will not pick up until he has ritually sacrificed his virgin daughter, Iphigenia. Considering whether to carry out this killing, Agamemnon changes his mind several times, but eventually does resolve to sacrifice his daughter. In some versions of the myth, Iphigenia is killed, while in others she is spirited away just before the fatal blow, a deer magically taking her place at the altar. In either case, the winds do pick up as prophesied and the Greek forces are able to sail for Troy.

Two premises incredible to modern audiences are taken for granted in the Greek myth: (1) the validity of the prophet's report (the story accepts as given that Iphigenia must be sacrificed before the Greeks will have sufficient wind to sail), and (2) the right of the father to sacrifice his daughter (the story question is whether he *should* sacrifice her, not whether he is allowed to).

Other human groups are subject to ritual sacrifice in ancient history and mythology—prisoners of war, for instance—but in the Iphigenia myth it is her status as Agamemnon's child that gives him the authority to decide whether she lives or dies. There is a gendered aspect to this story—a father is debating whether to kill his daughter—but it is not solely or primarily Iphigenia's sex that makes her available for sacrifice; it is her age, or, more precisely, her subordinate status as Agamemnon's child. Male children are also sacrificed in ancient legend, both in Greek mythology, and, most famously, in the biblical story of Yahweh ordering Abraham to sacrifice his son Isaac. The notion that the father may ritually kill any of his children is the ultimate logical consequence of a patriarchal family structure in which wives, children, servants, drafted animals, and "livestock" are understood as possessions of the male head of household.

What makes the Iphigenia myth useful for discerning relations between speciesism and intra-human hierarchies is the fact that the prototypical sacrificial victim is not a human child, but a nonhuman animal. What is being contemplated for Iphigenia is normally done to a sacrificial animal, and this comes out in the dramatic descriptions of preparations for her ritual slaughter, with the tragedians Aeschylus and Euripides naturally drawing comparisons with domestic animals such as goats and calves:

and her father,
after praying, though she clasped
his knees, begged him with all her heart,
ordered his men to lift her like
a goat, face downward, above the altar,
robes falling all around her, and
he had her mouth gagged, the bit yanked

roughly, stifling a cry that would
have brought a curse down on the house[4]

But for you, Iphigenia, there will be a different crown.
The curls of your hair the Greeks will adorn with a wreath,
Will take you like a brindled calf, pure, from a cave in the mountain rocks,
Will stain a human throat with blood[5]

Notwithstanding the shared subordination of children and animals within the patriarchal household, a father's sacrifice of his child is unusual and dubious, given the father's normal obligation to protect and provide for his children—an obligation that does not apply to his worked animals and living stock. In Colin Teevan's 1999 adaptation of Euripides' tragedy, entitled *IPH* ..., chorus members voice the problem:

CHORUS MEMBER: A father cannot kill his daughter –
CHORUS MEMBER: A child's not reared a calf for slaughter –[6]

Anxiety at the prospect of a *child* being sacrificed is indicated by the dramatic trope of the last-minute animal replacement, a common ending mentioned, for example, by Ovid in *Metamorphoses* ("Diana substituted—so they say—a deer for Iphigenia"[7]), and represented in *Iphigenia in Aulis* through a messenger speech:

MESSENGER: Suddenly, it was a miracle: everyone had heard the sound of the knife—but no one could see where in the world the young maid had disappeared to. The priest cried out. The army echoed his cry. And then they saw the miracle, impossible to believe even as it happened before their eyes. There on the ground lay a deer, gasping for breath. She was a full-grown deer, beautiful, and the altar of the goddess was dripping with her blood.[8]

The dramatic tension created by the prospect of child sacrifice is relieved by substituting a victim—an animal—of a type the audience is used to seeing sacrificed and, presumably, considers unproblematic. But, whether the child sacrifice is completed or narrowly averted, the dramatic representation of a child being bound to the altar in the manner of a domestic animal cannot help but raise questions regarding the patriarchal treatment of humans *and* animals. In his 1966 article on Greek tragedy and sacrificial ritual, Walter Burkert argues that the interchangeability of victim types, human and animal, calls into question the entire sacrificial system of Greek culture.[9] Burkert's abolitionist conclusion, however, may be forestalled simply by reinforcing the boundary between the human and nonhuman realms. Thus, in a recent commentary, Albert Henrichs defends animal sacrifice against Burkert's challenge by claiming that human sacrifice was always purely imaginary (no child was harmed in the making of this tragedy):

> In real life, animal victims did not turn into human victims, and human sacrifice remained strictly within the realm of the Greek *imaginaire*—it was one thing to think about human sacrifice, but it would have been unimaginably bad to put it into actual practice.[10]

Legendary depictions of human sacrifice blur the edges between the categories of human and nonhuman animals, an alarming implication for defenders of purely *human* rights, but one which may be, and typically is, handled by a reassertion of human sanctity—for example, by language such as Henrichs', according to whom cutting the throat of a child is "unimaginably bad," in some way that cutting the throat of a goat or a calf is not.

It is one thing to draw lines between abstract categories such as "human" and "animal," an easy exercise in semantics and one useful for asserting the anthropocentric status quo, but stories such as those told by the Greek tragedies draw us in narratively, inviting our identification with the characters, and thus have emotional consequences. If we identify with Iphigenia—and we are all invited to do so, whether male or female (recall that in classical Greek theater the authors, actors, and audience members were all male)—and Iphigenia is herself identified with sacrificial animals through her dire situation, then how do we not also identify with those sacrificial animals? Emotional identification with targeted animals in these stories is expectable, but is handled, in a culture intent on continuing its exploitative institutions, through an array of distancing devices which, if not already active within the play itself, will certainly be brought to bear as soon as one leaves the theater.

An analysis of these distancing devices is a subject in itself, treated elsewhere.[11] Here I focus on the potential for consciousness raising this identification carries. Ellen McLaughlin develops this potential in her trilogy *Iphigenia and Other Daughters*, a recent English-language adaptation of the Greek tragedies *Iphigenia in Aulis, Electra*, and *Iphigenia in Tauris*. In her version of *Iphigenia in Aulis*, McLaughlin brings out the emotion of Iphigenia's identification with a hunted deer, a member of the species traditionally said to be substituted for Iphigenia at the altar:

IPHIGENIA:
They killed a deer yesterday
I saw it while it was still running
Stamping down the boughs …
There I go, I thought, run fast
But they caught it
Of course
And when they brought it in
Limp and undignified, head lolling
I couldn't watch, I turned away
Because the eyes were open
And they saw me
Dead, they saw me
I shivered
You and me, I thought
We know each other[12]

McLaughlin develops this animal/female identification further in her *Iphigenia in Tauris*, the play that follows up on Iphigenia's life after she has been whisked away from the altar. At the beginning of the play, Iphigenia recalls what she saw looking down on the altar:

They say I was spirited away at the moment of the knife
Some deer died in my place
But it looked like me to me at any rate
It was disorienting
I rose up out of myself
Looking down at the figure on the stone
She's so young, I thought
Beautiful in a way[13]

Was an animal, one that looked very much like Iphigenia, sacrificed in her place as she was lifted bodily into another realm, or was the substitution just an exculpatory myth, and it is actually Iphigenia's released spirit that looks down on her own body? The ambiguity of this scene reinforces the interchangeability of girl and doe.

McLaughlin takes a further step in her version of *Electra*. The background to this play is that Clytemnestra killed her husband Agamemnon on his return from the Trojan War, a murder motivated in part by anger over his role in her daughter Iphigenia's sacrifice. Orestes, her son, has been in exile, but now returns in disguise, ambivalently intending to avenge his father's murder by killing Clytemnestra. To further his anonymity,

Orestes pretends to be a messenger bringing a story of Orestes' death to his family. In the following scene, the female/animal identification, already developed in the first part of the trilogy, allows us to comprehend Clytemnestra's response to Orestes' story.

ORESTES [in disguise]:   While leading a charge he [i.e., Orestes] fell from his horse and was dragged out into no-man's land. I crawled out that night to find him there tangled in the reins. It was a terrible death but heroic, lit by his valor.

(*Slight pause.*)
CLYTEMNESTRA: Sad. (*Assesses her own emotional state*) I'm taking it well, wouldn't you say? You look like a boy who knows his way around death and how people take it. My son. Huh. And his horse?

ORESTES (*Confused*):   I'm sorry?
CLYTEMNESTRA: What happened to the horse?
ORESTES: (*Somewhat baffled*) Well, the horse died, too.
CLYTEMNESTRA: Tragic. Dumb animals. The suffering of the innocent. Hardly seems fair.
ORESTES: Sorry?
CLYTEMNESTRA: Getting animals mixed up in such a loathsome thing. War. What do they know about it? What did they ever do to us? How confusing and terrifying it must be for a horse, a creature like that, all the noise, shouting of the dying, the cannons. Awful. (*She cries, genuinely moved*) Why should they suffer as we do? No justice. Excuse me.[14]

The implications of McLaughlin's rewriting are politically significant. By asking her son, "What happened to the horse?" McLaughlin's character Clytemnestra brings the well-being of the exploited animal, a horse drafted for war service, into the discussion, signaling its moral relevance. The fact that these kinds of questions are almost never asked is reflected by Orestes' incomprehension.

We, the audience, are likely also caught off guard. Within our narratives of human domination, the stories that depict people using animals and people using other people, animals serve a set function. Nonhuman animals appear as foils, there to define a naturalized realm of exploitation against which intra-human exploitation—the intended focus of the story—may be problematized. The point is not to similarly problematize our exploitation of animals, but to use the shared presumption of animal exploitation as background, an unproblematic given, to bring into relief the contrasting status of intra-human exploitation, which is challenged by the story or at least seen as challengeable.

Each time a story uses animals this way, it further naturalizes the human exploitation of animals. By rewriting the story so that Clytemnestra questions the use of horses in war, McLaughlin presents the human exploitation of nonhuman animals as cultural, not natural, and thus subject to moral assessment. In the conclusion of this chapter I return to this play, considering it as an example potentially pointing the way forward.

II. *The Taming of the Shrew*

The second narrative of human domination to be considered is William Shakespeare's *The Taming of the Shrew*, an early comedy (circa 1592), often considered one of Shakespeare's "problem" plays. This play does create problems, for directors, audiences, and critics, in that it presents for our amusement the spectacle of a man breaking the spirit of a woman. Regardless of how Shakespeare's original audience may have responded to the play, we do not find domestic violence and sadistic control so funny anymore—or at least we feel we should not find it funny.

Nonetheless, *The Taming of the Shrew* remains one of Shakespeare's most popular plays. A recent survey of North American Shakespeare productions reveals that *The Taming of the Shrew* is among Shakespeare's *most* frequently produced plays—only *Romeo and Juliet* and *A Midsummer Night's Dream* being staged more often.[15]

Have directors "solved" the play, finding ways to present it that do not make audiences feel like they're cheering for women's subordination? Or, does the play's continued success show that audiences, even today, enjoy voyeurizing heterosexual domination more than we'd like to admit? (Consider *Fifty Shades of Grey*.) In either case, Shakespeare's *Shrew* remains a significant literary representation of gender politics.

It also reveals deep connections between women's subordination and that of nonhuman animals. These connections are announced even before the curtain rises, within the play's title, which mentions animals twice, once directly and once indirectly. Directly, because a "shrew" is an animal; indirectly, because "taming" is most typically done to animals, though, in this case, it is done to a woman. We know a woman is targeted because the word "shrew," though originally an all-purpose insult applied indifferently to a man or a woman, had by Shakespeare's time come to be restricted to women, typically referring to a scolding, turbulent wife. But the primary reference of "shrew," of course, is to an animal, the transference of the term to humans occasioned, most likely, by naturalistic and/or superstitious perceptions of the small insectivore as malignant, venomous, and vicious.[16]

Two major sex/species associations are thus written into the title: what we (men) *do* to women and animals (attempt to tame them), and how we *think* of them (as having a common, animalistic, nature). Both these associations are borne out in the comedy itself. Petruchio, the fortune-hunting bachelor, enthusiastically takes up the challenge of subduing the notoriously "froward" Katherine when he hears of her substantial dowry, employing a variety of novel and amusing tactics toward breaking her will, so that (1) she will agree to marry him (though her actual consent is evidently of little concern to anyone in the play), and (2) she will obey him after their marriage. While the Bard evidently expected the hero's unusual tactics to delight and amuse audiences (given Petruchio's tactical originality compared to the usual, uninspired Elizabethan practices such as beating shrewish wives with a stick or mockingly parading them through town), modern audiences may be less impressed, having now become acquainted with all Petruchio's tricks via reports of, for instance, fraternity hazing rituals, military boot camp culture, the CIA's "enhanced" interrogation techniques, and, indeed, the routine controlling tactics of modern husbands and boyfriends. Petruchio was ahead of his time, but not ours.

What does he do to Katherine? He humiliates her in public, isolates her socially, withholds food and clothing while depriving her of sleep, and terrorizes her through violent displays, all carefully calculated to break down her sense of self-will and entirely replace it with his own. Much of this he does in explicit imitation of how certain nonhuman animals are bent to human purposes, as in the following speech, where Petruchio tells the audience how he treats Katherine like a falcon (also called a "haggard" and a "kite") being broken in:

> Thus have I politicly begun my reign,
> And 'tis my hope to end successfully.
> My falcon now is sharp, and passing empty,
> And till she stoop, she must not be full gorged,
> For then she never looks upon her lure.
> Another way I have to man my haggard,
> To make her come and know her keeper's call,

That is, to watch her, as we watch these kites,
That bate and beat and will not be obedient.
She eat no meat today, nor none shall eat.
Last night she slept not, nor tonight she shall not...
This is a way to kill a wife with kindness,
And thus I'll curb her mad and headstrong humor.[17]

The parallel drawn so strongly here between a wife and a domesticated animal is echoed throughout the play, strongly enough, in fact, that directors have added their own expression of the theme: a whip. This prop is not mentioned in the script itself, but has become conventional stage business, helping Petruchio show his inner nature ("for I am he am born to tame you, Kate"[18]), while delighting generations of audiences as he struts about the stage, menacing Kate by cracking the whip at her like a circus lion tamer.

As Petruchio's intended wife, Kate shares a common status with domesticated animals, falling within the same basic category of "things for a man to acquire." Petruchio woos Kate as one assesses a horse at auction, musing, "Why does the world report that Kate doth limp? ... Oh, let me see thee walk. Thou dost not halt."[19] Just after the wedding ceremony, in the course of arguing for his right to immediately take his wife away with him, against her wishes, before the wedding feast has even begun, Petruchio declaims the property status of women:

I will be master of what is mine own.
She is my goods, my chattels. She is my house,
My household-stuff, my field, my barn,
My horse, my ox, my ass, my anything.[20]

This outburst, extreme even by Elizabethan standards, is probably meant to provoke audiences, rather than convince them. But, is it sufficiently farcical to deflate the pretensions of male dominance? Of course, one is free to so construe it, but there is precious little in the text of the *Shrew*—or in Shakespeare's wider canon, for that matter—to support an image of the poet as any sort of closet proto-feminist.

Nonetheless, *The Taming of the Shrew* remains a perennial favorite, a true theatrical hit, and as such, theater folk have long sought ways to live with the play—that is, to live with themselves while producing the play. John Fletcher, Shakespeare's successor as company playwright, apparently already feeling uncomfortable with Petruchio's total victory over

Katherine, wrote a sequel to his colleague's hit called *The Tamer Tamed*, in which a remarried Petruchio more than meets his match in his new wife. Just knowing that Fletcher's play exists, that somewhere out there floating around in Plato's abstract realm of forms Petruchio is receiving his come-uppance, has allowed one director, Gregory Doran, to justify producing what he calls Shakespeare's "gruesome" and "unpalatable" spectacle of Kate's subjugation.[21]

Doran played it straight, allowing the *The Shrew*'s inherent misogyny to speak for itself, secure in the knowledge that the "antidote" is already out there in the form of Fletcher's sequel. Many other directors and per-formers, though, seek ways, more or less subtle, to subvert the play's patriarchal message while remaining true to the letter, if not the spirit, of Shakespeare's script.

Most of these subversive efforts focus either on the end of the play, Katherine's famous "thy husband is thy lord" speech, or on its beginning, the "induction" or framing story. Katherine's 44-line speech, affirming the wife's subordination to her husband—"such duty as the subject owes the prince"—is delivered at the behest of her newly in-charge husband as he seeks to win a wager with his buddies over who has the most compliant bride. Drawing the play to its close, this depressing speech seriously jeop-ardizes any production's chance of sending the comedy's audience away feeling merry or uplifted. The temptation to ironize or in some way bracket the speech is thus great, even though there is no indication in the script that the speech is meant as anything less than fully sincere. A famously extreme example of the ironizing tactic is seen in the first Shakespearean "talkie," the 1929 United Artists film starring Mary Pickford and Douglas Fairbanks, Jr., in which Pickford concludes her rendition of Katherine's long speech by delivering an enormous wink to the film's audience, letting us know that *this* Shrew has not been tamed, appearances notwithstanding.

Another, nearly opposite, strategy is Doran's—not to ironize the speech but to emphasize the change it indicates in Kate's character, from a difficult, but energetic, woman choosing her own path to a crumpled shell now only capable of speaking in her husband's voice. Under this interpretation, *The Taming of the Shrew* functions not so much as comedy but as a consciousness-raising event: this is what domestic abuse looks like. A salutary work, if no longer very funny.

The other major subversive approach is to make full use of the induc-tion, a two-scene prelude which sets the story of Petruchio and Katherine within a wider, framing narrative. The induction tells the tale of a young lord and his servants coming on a tinker, Christopher Sly, passed out drunk. The lord decides it would be great fun to convince Sly, once he awakens, that he is the lord of the manor, all his memories to the contrary merely signs of a long-suffered mental illness. The (actual) lord drafts his servants into the deception, going so far as to dress his page, Bartholomew, as a woman to pass him off as Sly's wife. When a company of players happens by and offers their services, the young lord jumps on this as another way to convince Sly of his elevated status by presenting a play to him, private performances by traveling troops being one of the perquisites of the well-to-do nobility. Thus, the story of Petruchio and Katherine is, in Shakespeare's original script, a play within a play, performed to convince a clueless buffoon that he commands more than he actually does.

These induction scenes are often cut, but retaining them has the immediate effect of bracketing the female-taming story—we, the audience, need no longer just consume the tale, but are invited to step back and consider its implications as a work of fiction. The details of the framing story are significant in suggesting how we may understand Petruchio's campaign to break Kate as another elaborate game of role-playing. It is not just that Sly is given an inflated sense of his own power; additionally, a specifically gendered component to the deception emerges through the process of disguising the page as his wife. The lord imparts careful instructions regarding how Bartholomew should act as a wife:

> Such duty to the drunkard let him do
> With soft low tongue and lowly courtesy,
> And say, 'What is't your honour will command,
> Wherein your lady and your humble wife
> May show her duty and make known her love?'[22]

This scene cannot but suggest to the attentive reader or audience member the possibility that gender roles in general—Petruchio's and Kate's, as well as those we inhabit in real life—are fictional characters created via the lines we deliver and our actions "on stage." This insight would then strongly suggest that Katherine's concluding speech of abject submission is likely also just part of the act.

One of the reasons the induction is often omitted is that, inexplicably, Shakespeare's script does not bring the framing story back at the end of the play, a dramatically and aesthetically unsatisfying imbalance that can be fixed by cutting the induction altogether. Directors determined to make use of the induction's subversive implications, however, have not let the absence of lines and stage directions for Sly deter them from rounding off the play: they just bring the framing story back silently, without lines. A Cincinnati Shakespeare Company production I attended recently, for example, concluded with the actors moving upstage to begin removing their costumes, reminding us that the Petruchio/Kate tale is a play-within-a-play. While this was going on, the actress playing Kate came downstage, by herself, and stowed away in a trunk the "sweet clothes" used to convince the tinker Sly of his noble status, her action a potent symbol of the inessential and removable nature of men's power. Similarly pointed in its reassertion of the frame, the 1998 American Repertory Theatre production, directed by Andrei Serban, ended with the actors familiarly embracing each other, as if already relaxing offstage, then walking off hand-in-hand in various couples, some same-sex, some mixed, thus visually representing the multiplicity of gender roles available in a world moving beyond predatory heterosexuality.

Though critics and directors have become savvy regarding the framing story's potential for destabilizing Petruchio's marital coronation, they have not considered the connections it makes with the domination of nonhuman animals. As in the main storyline, where associations made between the to-be-tamed Katherine and various animals (such as the shrew of the title or the falcon of Petruchio's strategy speech) are rarely remarked on or analyzed, so also, in the induction, connections between the subordination of women and animals are largely ignored.

But, those connections are present. A hunting theme is very prominent from the beginning of the induction, for instance, and Shakespeare's script relates it to Petruchio/Katherine and gender relations more generally in a number of ways. For instance, among the offerings deployed to convince Sly of his noble status are horses, hounds, and hawks, the living instruments used in sport hunting:

> LORD: Or wilt thou ride? Thy horses shall be trapped, ...
> Dost thou love hawking? Thou hast hawks will soar
> Above the morning lark. Or wilt thou hunt?
> Thy hounds shall make the welkin answer them.[23]

The association between women and these domesticated animals is apparent because a wife is similarly made available to Sly, the faux noble, and, like the horse, hawk, and hound, she is expected to obediently serve his wishes (this assumed obligation justifying whatever means might be necessary for correcting the wife, or falcon, who "bates, beats, and will not be obedient").

The common station of women and household animals is also high-lighted by a suggestive, though heretofore unremarked on, parallel between the opening and closing scenes of the play. In the closing scene, the three new husbands good-naturedly vie with each other over who has the most compliant wife. In the opening scene, the young lord banters with his first huntsman over which of the hunting dogs is best:

> LORD: Saw'st thou not, boy, how Silver made it good
> At the hedge corner, in the coldest fault?
> I would not lose the dog for twenty pound.
> FIRST HUNTSMAN: Why, Belman is as good as he, my lord, ...
> Trust me, I take him for the better dog.[24]

This correspondence between wives and hounds—each living possessions of the master, suitable for ranking as to who serves him best—might easily be lost with the entirety of the play coming between the two scenes, but Shakespeare helpfully gives Petruchio a line in the closing scene which makes the connection explicit: "Twenty crowns! I'll venture so much of my hawk or hound, but twenty times so much upon my wife."[25]

The dramaturgical work done over the last century to find ways to present Shakespeare's *Shrew* without condoning spousal abuse has not been accompanied by any similar work to counter the animal abuse also intrinsic to the play. On the rare occasions when the connection between female and animal subordination is noted, the tendency is to bemoan how unfortunate it is that women are sometimes treated like animals. Thus, Shirley Nelson Garner criticizes the abuse of Kate: "Petruchio's physical taming of Kate is objectionable in itself; it is particularly humiliating because it is 'appropriate' for animals, not people."[26] Garner's use of scare quotes for the word 'appropriate' indicates she is not quite prepared to concede that the corresponding physical taming of nonhuman animals is morally acceptable. Since there is no follow-up to her implied point, however, the question of how animals *should* be treated is left unaddressed, thus reinforcing the status quo.

The fact that Petruchio treats Kate as property, like his cattle, that he wagers with other men over her excellence, as he does over his hunting hounds, that he breaks her will by depriving her of food and sleep, as with his falcons, is now considered abusive, all the more so because it humiliates his wife by lowering her to the status of an animal. The anti-sexist (but not anti-speciesist) redress is clear—stop treating women like that. And if you must present *The Taming of the Shrew*, find a way to stage it so the audience knows that Petruchio's perspective on marriage is not the only one. This is currently the received view in the theatrical world. As yet unoffered is a production in which the audience sees that Petruchio's perspective on animal abuse is also not the only one available to us.

III. Uncle Tom's Cabin

The two plays analyzed here, *Iphigenia in Aulis* and *The Taming of the Shrew*, are nonpolitical works, in a sense. Though their subject matter reveals connections between, in the first case, the subordination of animals and children, in the second, animals and women, neither work takes as its primary purpose the promotion of any particular social cause. For my next examples, I turn to two overtly political works, novels self-consciously written to further social justice.

Harriet Beecher Stowe wrote her 1852 novel, *Uncle Tom's Cabin: Or, Life Among the Lowly*, in response to the passage of fugitive slave laws, laws requiring those living in northern states to turn in fugitive slaves so they could be returned to their owners in the south. The moral outrage at being drawn into active participation with the "peculiar institution" goaded Stowe into writing her abolitionist novel. *Uncle Tom's Cabin* was hugely successful, becoming the best-selling novel of the nineteenth century and raising consciousness both in the USA and abroad regarding the evils of slavery.

Several themes underwrite the book's abolitionist position: (1) all people have immortal souls, including the people of color who are enslaved, (2) enslaved people of color feel their oppression deeply, especially the trauma of being separated from their spouses or children when they are sold off, (3) people of color can be of high character and act in virtuous ways even though enslaved, and (4) enslaved people can also be degraded by their condition, becoming brutalized from the unrelenting oppression of being used purely to increase another's wealth.

A quick way to summarize the effectiveness of *Uncle Tom's Cabin* would be to say that it "humanizes" those who have been enslaved. While this label is accurate so far as it goes, it becomes problematic if animal liberationists seek to emulate that novel's approach. It seems obviously a contradiction in terms, to "humanize" nonhuman animals, but even if not logically self-defeating, the strategy may still be ill-advised, for example, by prompting us to project onto nonhuman species characteristics that are not present, applicable, or relevant.

Stowe's basic strategy of opposing slavery by affirming the full humanity (spiritual, moral, and sensible) of enslaved people of color may transfer well to other intra-human causes, while being useless or counterproductive within the animal rights movement. Consider the premise regarding immortal souls. Stowe wrote from a deeply Christian perspective, arguing fervently against slavery on the grounds that enslaved people of color possessed immortal souls, "bought with blood and anguish by the Son of God."[27] Regardless of the actual extent to which this premise was accepted during Stowe's time, she wrote as if it were obvious and incontestable, and not a single character in *Uncle Tom's Cabin* challenges it. The same incontestability cannot be assumed regarding the souls of nonhuman animals. I have served in various Methodist churches for the last 15 years, and during that time entered numerous discussions with people, both inside and outside of the church, regarding the fate of animals after death. The view that there will be animals in heaven is common, both among clergy and laity, but it derives not from a doctrine of animal souls. Rather, heaven being understood as a place of perfect reward for saved humanity, many conclude that animals must, of course, be there—for how could they, the faithfully devoted, experience perfect, heavenly bliss without the companionship of their beloved pets? From such discussions, one gets the picture of an immaculately landscaped golden avenue populated by cheerful neighbors, well-behaved cats and dogs, and songbirds in the treetops (off-limits to the cats?), but certainly no predatory mountain lions, venomous snakes, or disease-carrying rats in the neighborhood, or, God forbid, any gnats, ticks, or fleas.

The attribution of immortal souls to all animals, or even, say, just to all mammals (including those that many people eat, such as pigs and cows?), would certainly not gain the widespread agreement that Stowe presumes with regard to the souls of black folks. This may not be a tremendous loss for animal liberation strategy though, given that the immortal souls argument is likely not the most moving part of Stowe's narrative. There is a logical gap from this premise to the abolitionist conclusion: why does the possession of a soul make the person's enslavement wrong? Not only is this gap never closed within the book, the weakness of the argument is tacitly acknowledged through the behavior of the story's slaveholders, who accept that their human chattel have immortal souls but go about their business unperturbed; if anything, they are encouraged by the premise, evidently figuring that the "good" slave's eternal future reward will more than compensate for his or her relatively few years of appropriated labor while on earth.

Stowe's second point, her insistence on the enslaved person's capacity for virtue and moral excellence could be, and has been, matched by similar arguments regarding the virtues of nonhuman animals. A number of recent books explore the reality of nonhuman animals as heroes or exhibiting other sorts of positive moral agency.[28] Such stories and studies can counter the stereotyped images, self-servingly developed around the various exploitative industries, of animals as inherently amoral or vicious ("pig" as a synonym for greed, "sheep" for passive subservience, "wolf" for sexual aggression, etc.).

How far recounting stories of animal virtue may go in encouraging liberationist perspectives is worth exploring. Could anyone allow medical experimentation on a dog who had pulled a family member out of a burning building? Or countenance the commercial slaughter of dolphins, knowing how they go out of their way to assist people who are drowning in the ocean? The fact that the virtuous animal approach is unlikely to be applicable to every species subjected to human domination should not dissuade us from making use of it when it does apply.

One lesson from the reception history of *Uncle Tom's Cabin* should be mentioned here. The title character of the book, Uncle Tom, is Stowe's most thorough attempt at creating a character who, though enslaved, exemplifies moral virtue to the highest degree. Her Uncle Tom is pious, honest, diligent, unfailingly kind and patient, refusing to speak ill of or raise his hand against even the most oppressive masters—the sort of character Stowe may have developed by asking herself, "How would Jesus have acted if enslaved?" Notwithstanding Stowe's evident intention for Uncle Tom to stand as a paragon of virtue under enforced servitude, the received meaning of the epithet "Uncle Tom" reflects a rather different character. According to the online Urban Dictionary, for example, the primary meaning of "Uncle Tom" currently is this: "A black man who will do anything to stay in good standing with 'the white man,' including betray his own people." This accurately accords with my sense of the term, as I have come across it any number of times, such as, to cite one example, historian C. Vann Woodward's statement, "There was a certain amount of fawning Uncle-Tomism among the Negroes."[29]

The standard meaning of the phrase "Uncle Tom" does not just differ from Stowe's authorial intent, it actually contradicts the behavior of the character as he is depicted in the very book that originated the term. The Uncle Tom of Stowe's book explicitly rejects "white man's law" when it conflicts with what he takes to be a higher law, that of the Bible. In fact, in two key incidents near the end of the book, Tom deliberately and openly rejects the course of staying "in good standing with the white man": first, he refuses to whip other slaves, and later, in a deliberate act of conscientious disobedience that brought him a fatal, and foreseeable, beating, he refuses to help track down two runaway slaves (both women whom the plantation owner, Simon Legree, kept for sexual service).

This is a striking example of a chastening lesson: writers do not have full control over the meanings we try to create. We can only try, and try again.

One of Stowe's main abolitionist strategies is to render her enslaved characters as people with strong social attachments. She deliberately contradicts the self-serving view of the white owning class that women and men of color recover quickly and easily when torn apart from their spouses and children. Rather than saying Stow "humanizes" these enslaved individuals, for us it is more useful to say that she "subjectivizes" them, that is, portrays them as subjects of their own lives, with desires, dreams, preferences, reactions, all the qualities of having experiences and an inner life. Stowe's point about enslaved people can be carried over into the animal world. I remember first hearing about how cows, both individually and as a herd, react with agitation when a calf is taken from the mother (as typically must be done so that the milk the postpartum cows are producing can be collected and sold). I was just beginning to explore animal rights at the time, and the image of cows lowing in protest and grief over the loss of their young, an image confirmed through multiple conversations with dairy farmers ("yeah, they always hate it"), helped motivate my nascent veganism and activism.

Part of what is so moving in the image of deep maternal attachment between cows and their calves is that it is unexpected—it makes them seem so much "like us." The background to this sense of surprise is the ideology of nonhuman "brutishness." The reason we have a single word, *brute*, meaning both "nonhuman animal" and "a creature with limited capacity for feeling," is because this conflation furthers the process of animal exploitation, encouraging our placating belief that farmed animals do not suffer deeply from what is done to them. Stowe placed the real subjective experiences of enslaved human beings at the center of her narrative; we may write similarly, telling stories of nonhuman animals and their felt experiences, both in and out of the various institutions of exploitation.

The fact that some of Stowe's anti-racist strategies may work analogously for animals does not mean, unfortunately, that *Uncle Tom's Cabin* and other anti-racist narratives are unalloyed comrades of the animal rights movement. The problem is that Stowe does not limit her advocacy to affirming the spirituality, morality, and subjectivity of slaves. As do most progressives, in making her case for human liberation, she reflexively affirms an opposition between those humans and other animals, as if a useful, if not obligatory, way to free people of color is to assure us that they are, in fact, not animals. This logic, of course, implies that enslaving or otherwise exploiting nonhuman animals is natural and appropriate. A few examples follow.

Haley, a callous and amoral slave trader, compares Negroes to other, nonhuman, living property: "they is raised as easy as any kind of critter there is going; they an't a bit more trouble than pups."[30] Derogatory slang terms such as "coon" (short for "raccoon") are used by the servilely sadistic Sambo, himself a black slave: "Mas'r, let me lone for dat ... I'll tree de coon. Ho, ho, ho!"[31] By putting such lines in the mouths of unsympathetic characters like Haley and Sambo, Stowe expresses her disapproval of the common practice of associating Negroes with nonhumans.

Stowe's uneasiness with people of color being thought of, or treated like, animals is also indicated through the statements of her sympathetic characters. The always respectful and thoughtful youngster George Shelby (who grows up to free his slaves at the end of the novel), chides a slave trader in the following terms: "I should think you'd be ashamed to spend all your life buying men and women, and chaining them, like cattle! I should think you'd feel mean!"[32] Another white man who frees his slaves, the "honest old John Van Trompe," reacts on hearing of the fugitive Eliza, fleeing her owner to prevent him from selling off her young son: "That's natur now, poor critter! hunted down now like a deer,– hunted down, jest for havin' natural feelin's, and doin' what no kind o' mother could help a doin'!"[33] Another white character, referred to as "the honest drover," also speaks strongly (if prudentially) against treating enslaved people as animals, saying, "Treat 'em like dogs, and you'll have dogs' works and dogs' actions. Treat 'em like men, and you'll have men's works," and, "The Lord made 'em men, and it's a hard squeeze getting 'em down into beasts."[34]

Stowe's narrator also disapproves of enslaved people being treated as we routinely treat nonhuman animals, as in this description of a slave auction: "Various spectators ... gathered around the group, handling, examining, and commenting on their various points and faces with the same freedom that a set of jockeys discuss the merits of a horse."[35] The suggestion that what counts as unjust mistreatment, when meted out to a human being, is perfectly fine, normal, even "happy," when done to an animal, is made explicit through Stowe's story of Tom Loker. Having been shot in the leg while in mercenary pursuit of runaway slaves, Loker is nursed back to health by abolitionist Quakers, and leaves them a changed man:

> Tom arose from his bed a somewhat sadder and wiser man; and, in place of slave-catching, betook himself to life in one of the new settlements, where his talents developed themselves more happily in trapping bears, wolves, and other inhabitants of the forest, in which he made himself quite a name in the land.[36]

The ideology of human chauvinism lay behind these statements of Stowe's characters and her narrator. Race-based slavery is wrong, according to this perspective, because it is inappropriate for human beings, who are by nature "higher" than the nonhuman animals we may "happily" trap in the forest or domesticate and sell at market. This hierarchical vision of (male) humanity over the animal is behind Stowe's description of the two debased slaves, Sambo and Quimbo, who, in her words, "seemed an apt illustration of the fact that brutal men are lower even than animals."[37] Thus, the narrator explains what freedom means to George Harris, who, in the course of the novel, frees himself from slavery, gains education, lives for a time in France, and finally emigrates to Liberia to start a new life with his family: "To him, [freedom] is the right of a man to be a man, and not a brute."[38]

From an anthropocentric standpoint, slavery is wrong because it unnaturally and illegitimately lowers a (male) human to the status of an animal; abolition of slavery is required so that a black man can be a man, not a nonhuman animal, or "brute." This oppositional strategy—raise up (male) people of color by distancing them from nonhuman animals—has two simultaneous consequences: it challenges the exploitation of those people while confirming the institutionalized abuse of animals.

The oppositional strategy is ubiquitous across the literature of human protest movements. It is difficult to find examples of progressive writing that do not, at some point, reinforce speciesist exploitation. To cite just one recent example, consider this passage from Michelle Alexander's book, *The New Jim Crow*:

> The stigma of race has become the stigma of criminality. Throughout the criminal justice system, as well as in our schools and public spaces, young + black + male is equated with reasonable suspicion, justifying the arrest, interrogation, search, and detention of thousands of African Americans every year ... One student complained ..., "We can be perfect,

perfect, doing everything right and still they treat us like dogs. No, worse than dogs, because criminals are treated worse than dogs."[39]

The logic of human domination has not changed since Stowe's time. The use and abuse of nonhuman animals is still the default model for describing the oppression of a human group, such as, in this case, that of young black men racially targeted by police in America's current age of mass incarceration. Every time a speaker or writer uses animal exploitation as a model for human injustice (or, as in Alexander's book, presents such a quote without comment, as if the analogy is unproblematic), he or she reinscribes human supremacy, reinforcing the dogma that the exploitation of animals by people is, in some deep but unnamed sense, natural.

It is important to see that Stowe's use of the oppositional strategy is gratuitous. She could have presented all of her main arguments against slavery—spiritual, moral, and subjective—without ever bringing in comparisons with nonhuman animals. Consider, for instance, her description of Uncle Tom's reaction as he observes the devastation of Lucy, a young woman whose child was taken from her and sold, never to be returned to her again:

> His very soul bled within him for what seemed to him the *wrongs* of the poor suffering thing that lay like a crushed reed on the boxes; the feeling, living, bleeding, yet immortal *thing*, which American state law cooly classes with the bundles, and bales, and boxes, among which she is lying.[40]

Here, the oppressed slave is contrasted not with nonhuman animals, but with inanimate objects, "bundles, and bales, and boxes." Stowe's design is to elicit compassion for the mother's suffering while stoking anger at the legal system that allows it. This direct, nonoppositional approach to describing harm does no damage to the status of animals; in fact, it may incrementally improve their standing, since animals live and have feelings also, and any moral argument is, by its nature, generalizable.

Indeed, the common ground between human beings and other animals, each of us a subject of our own lived experience, is recognized in Stowe's text, for example, in the following quote from the backwoodsman Phineas:

> "Yes," said Phineas, "killing is an ugly operation, any way they'll fix it,–man or beast. I've been a great hunter, in my day, and I tell thee I've seen a buck that was shot down, and a dying, look that way on a feller with his eye, that it reely most made a feller feel wicked for killing on him; and human

creatures is a more serious consideration yet, bein', as thy wife says, that the judgment comes to 'em after death."[41]

The doctrine of human immortality is beside the point, but we may understand it being brought in here—at the last second, as it were—as a means of salvaging the anthropocentric ethic threatened by the hunter's knowledge of what it feels like to kill a buck. But the honest statement of the experience stands, pointing us toward a rhetoric that might challenge racist exploitation as strongly as does *Uncle Tom's Cabin*, but without condemning nonhuman animals in the process.

IV. The Jungle

When Upton Sinclair set out to write the great American socialist novel, he took Harriet Beecher Stowe's anti-slavery novel as his model. In response to the breaking of the Chicago meatpackers' strike of 1904, Sinclair had written a scathing anti-capitalist article for the Socialist newspaper *Appeal to Reason*. His piece so impressed the editors they suggested he write a longer exposé of the conditions workers faced under industrial capitalism. He agreed, taking a five-hundred dollar advance so he could move into the stockyards and observe working conditions firsthand. *The Jungle* was the result of that research.

Sinclair hoped *The Jungle* would galvanize support for socialism the way that *Uncle Tom's Cabin* had boosted abolitionism. Prior to its serialization in the *Appeal to Reason*, Sinclair wrote the following précis of his project:

> It will set forth the breaking of human hearts by a system which exploits the labor of men and women for profits. It will shake the popular heart and blow the top off of the industrial tea-kettle. What Socialism there will be in this book, will, of course, be imminent; it will be revealed by incidents—there will be no sermons. The novel will not have any superficial resemblance to "Uncle Tom's Cabin." Fundamentally, it will be identical with it—or try to be.[42]

While *The Jungle* did spark a social movement, it was not the one Sinclair intended. As an immediate consequence of his revelations regarding the shortcuts routinely taken within the slaughterhouses, new regulatory legislation was passed, evidently designed to assure disgusted and frightened consumers that their meat supply would be made safe. Nothing was done, however, to improve the wages or working conditions for the laborer. As Sinclair wryly noted, "I aimed at the public's heart, and by accident I hit it in the stomach."[43]

A further irony with *The Jungle* lies in its relationship to the treatment of cattle, pigs, and sheep, the so-called livestock (better, "living stock"). One might think that a novel exposing conditions inside the slaughterhouse would raise concerns for the slaughtered animals themselves, perhaps leading to a movement for humane slaughter procedures, if not a collective boycott of meat and the shuttering of slaughterhouses altogether. This has not been the case. *The Jungle* is still widely read, often assigned within the US high school curriculum, but moral and political reactions, both pro and con, focus on the exploitation of workers and on Sinclair's socialist message. The treatment of the animals is a tangential and purely prudential concern, an issue of whether slaughterhouse procedures produce a clean and safe food supply, and whether things have "gotten better" since Sinclair's time. The book does seem to have temporary effects on some individual readers' diets, producing brief spates of quasi-vegetarianism, but the effect comes about through disgust, not moral or compassionate considerations, as one can see by perusing online reviews at Web sites such as Goodreads.com. Comments such as, "[don't] discuss *The Jungle* extensively in your junior year literature class directly before lunchtime on hot dog day," and "I found, like most do, the narrative concerning the food production in the packing industry extremely distasteful. I actually had to avoid reading this book during my lunch hour," are the norm.

Considering that it is a book about the slaughter industry, *The Jungle* includes surprisingly few descriptions of what is done to the slaughtered animals; only one section covers this, in fact. This is not to say that animals are absent or insignificant. Sinclair references nonhuman animals throughout the book, but nearly always as ways to make points about human workers or the owning class. He uses animals instrumentally, employing them as symbols and foils, not as real beings with experiences that matter to them. He uses animals as a convenient stockpile of images to fill out his narrative descriptions, his rhetorical exploitation of them a semantic reflection of the exploitation occurring on animal bodies in the stockyards.

Sinclair's most frequent use of animal imagery is as shorthand descriptors—stereotypes and clichés—to designate human character, behavior, and status. Three stereotypes recur: animals as inarticulate, animals as passionate, and animals as heteronomous.

Animal imagery is used to describe humans who make sound without speaking, either because they are too young to talk—"when she put [the baby] into the basin he sat ... squealing like a little pig"—or because they are exhausted or overwhelmed with emotion—"he would fling himself into [the snowdrifts], plunging like a wounded buffalo, puffing and snorting with rage," and, "he stood there, ... the veins standing out purple in his face, roaring in the voice of a wild beast, frantic, incoherent, maniacal."[44] Sinclair also draws on animal-based clichés to represent instinctive and/or ungovernable drives and desires, such as aggression, mother-love, lust, or greed:

> He fought like a tiger, writhing and twisting. ... [Later], he was like a wild beast that has glutted itself.

> Human beings writhed and fought and fell upon each other like wolves in a pit.

> She had to bury one of her children—but then she had done it three times before, and each time risen up and gone back to take up the battle for the rest. Elzbieta was one of the primitive creatures: like the angle-worm, which goes on living though cut in half; like a hen, which, deprived of her chickens one by one, will mother the last that is left her.

> The whole of society is in [the owners'] grip, the whole labor of the world lies at their mercy—and like fierce wolves they rend and destroy, like ravening vultures they devour and tear![45]

Finally, Sinclair uses animal imagery to describe people falling under the sway of others and being abused or exploited:

> Lithuanians and Slovaks and such, who could not understand what was said to them, the bosses were wont to kick about the place like so many dogs. They were like rats in a trap, that was the truth; and more of them were piling in every day.

> Two years he had been yoked like a horse to a half-ton truck in Durham's dark cellars, with never a rest ..., and with never a word of thanks—only kicks and blows and curses, such as no decent dog would have stood.

> The sausage room was an interesting place to visit, for two or three minutes ... well-dressed ladies and gentlemen [touring the slaughterhouse] ... came to stare at [the worker], as at some wild beast in a menagerie.

The vast majority [of the strike-breakers brought in] were "green" negroes from the cotton districts of the far South, and they were herded into the packing-plants like sheep.[46]

The cumulative effect of Sinclair using animal imagery to describe people when they are inarticulate, passionate, or exploited is to create an image of nonhuman animals as essentially inarticulate, instinctive, and available to be preyed upon.

The problem, as seen in this novel, is the same as that expressed in *Uncle Tom's Cabin* and other anti-racist literature: not that we treat animals badly, but that we sometimes treat human beings just as badly or worse, as if they were animals:

They had put him behind bars, as if he had been a wild beast, a thing without sense or reason, without rights, without affections, without feelings. Nay, they would not even have treated a beast as they had treated him![47]

For a human male to be treated as we routinely treat nonhuman animals, or "beasts," is to degrade him in status, making him contemptible in the eyes of other men. Thus, when the protagonist Jurgis is jailed for attacking the supervisor who seduced his wife, he shares a cell with a professional criminal, Jack Duane, who, after getting to know Jurgis, develops, "an amused contempt for Jurgis, as a sort of working mule."[48]

Both the narrator and the characters of *The Jungle* denounce the exploitation of stockyard workers as a sort of category error, mistakenly treating (male) human beings as if they were disposable by nature, like nonhuman animals:

He was of no consequence—he was flung aside, like a bit of trash, the carcass of some animal.

"I couldn't keep ye after November—I ain't got a big enough place for that."
"I see," said the other, "that's what I thought. When you get through working your horses this fall, will you turn them out in the snow?" (Jurgis was beginning to think for himself nowadays).

And the [superintendent, railing at striking workers] shook his fist at them, and shouted, "You went out of here like cattle, and like cattle you'll come back!"[49]

Sinclair concludes his novel with a lengthy speech (notwithstanding his promise of "no sermons") by a fiery socialist organizer, whose promise of salvation to working men is couched precisely in terms of lifting them out of their animal-like status, affirming that they need no longer "plod on like beasts of burden, thinking only of the day and its pain."[50]

Though Sinclair makes it clear that women are also exploited on the factory floor, and at least two different women are, in passing, mentioned as socialists, the radical labor movement described in the book is organized by men and directed toward men. Women's problematic place within it connects with the author's perception of women's nature as ambiguously situated: somewhat human, somewhat animal. Sinclair's female characters are instinctive, highly emotional, apolitical, and plodding—the very qualities attributed to animals and seen as making them exploitable by nature. While most of characterizations of working men as "beasts of burden" are done oppositionally, problematizing their degraded status, Sinclair slides into an almost admiring equation of women with draft animals, as in the following description of Marija's work life:

> She was shut up in one of the rooms where people seldom saw the daylight; ... she stood on an ice-cold floor, while her head was often so hot she could scarcely breathe ... But because Marija was a human horse she merely laughed and went at it.[51]

The deep problem women pose for Sinclair is sexual. His protagonist Jurgis is only able to escape the trap of selling his labor for subsistence wages when his wife and children have died; at that point, he leaves Chicago to take up a hand-to-mouth existence in the fields and farmlands of the country. This preferred life has the great advantage of being free, and, as Sinclair portrays it, only two forces are sufficient to draw professional tramps back into the confinement of wage labor: cold winters, and the periodic lure of a "spree" in the taverns, where a man with money in his pockets can buy liquor and women. Women are thus seen as pulling men into the degradation of capitalist servitude in two ways, as wives (trapping them with children and the responsibility to provide), and as prostitutes (access to whom requires a supply of cash).

One might naturally suppose that the book's titular "jungle" metaphor refers either to the violence between humans and slaughtered animals, or to the Darwinian struggle between the classes the socialist author seeks to expose. Surprisingly, though, the only time the word "jungle" appears in the text, it refers neither to human/animal nor to owner/worker relations; it is used to describe a sexual encounter between Jurgis and a saloon girl during his hobo days:

> There was laughter and singing and good cheer; and then out of the rear part of the saloon a girl's face … she came and sat by him, and they had more drink, and then he went upstairs into a room with her, and the wild beast rose up within him and screamed, as it has screamed in the jungle from the dawn of time.[52]

The exploitation of working men by capitalist owners, the central injustice *The Jungle* seeks to address, is positioned by Sinclair within a larger framework of subjects: men, women, and nonhuman animals, each understood as possessing a distinct essential nature. The revolutionary goal is to achieve autonomous manhood for all human males, rejecting the ungoverned passions associated with animals—aggression, greed, lust—by forming networks of socialists, largely male groups within which workers can learn discipline (how to tame their animalistic passions for violence and sex) so they may work together to ultimately dispossess the profit-driven owners (thus removing rapacious greed from the economic system).

This socialist praxis, whatever its prospects may be for the empowerment of working class men, necessarily retards progress for women and animals: women, because they are associated with animals, the very natures autonomous manhood must oppose in itself and control in others; and animals, because their status as the essences paradigmatically fated for human exploitation is reinscribed through the novel's rhetoric.

Speciesist rhetoric is not essential to Sinclair's socialist goals, no more than it was to Stowe's abolitionism. Various passages in *The Jungle* demonstrate that effective criticism of capitalist exploitation does not depend on opposing the human to the animal. The utter expendability of the worker under unbridled capitalism is, for example, pointed out simply and directly in this line, about the fate of a working man once he is injured: "That he was utterly helpless, and had no means of keeping himself alive in the meantime, was something which did not concern the hospital authorities, nor any one else in the city."[53] The condemnation can be voiced more histrionically, in the style of the socialist organizer's speech, without yet bringing in any human/animal disjunction:

> I speak with the voice of the millions who are voiceless! Of them that are
> oppressed and have no comforter! Of the disinherited of life, for whom there
> is no respite and no deliverance, to whom the world is a prison, a dungeon
> of torture, a tomb! ... With the voice of those, whoever and wherever they
> may be, who are caught beneath the wheels of the juggernaut of Greed![54]

When expressed directly in terms of harmfulness, without the import of
extraneous speciesist clichés, the critique becomes transferable to other
species, the generalizable oppressed, "whoever and wherever they may
be." This is an especially important observation in the context of *The
Jungle*, since the author does include the one section describing the harms
done to animals as they are slaughtered.

By including a description of the slaughter process, Sinclair opens the
door to a sympathetic response to the animals by his readers—or, indeed,
by his own characters, such as Teta Elzbieta in this passage:

> Jokubas pointed out the place where cattle were driven to be weighed ... All
> night long this had been going on, and now the pens were full; by to-night
> they would all be empty, and the same thing would be done again.
> "And what will become of all these creatures?" cried Teta Elzbieta.
> "By to-night," Jokubas answered, "they will all be killed and cut up, and
> over there on the other side of the packing-houses are more railroad tracks,
> where the cars come to take them away."[55]

Sinclair allows his narrator to comment on the inherent injustice of the
"very river of death" to which the thousands of unsuspecting hogs are led:

> [T]he most matter-of-fact person could not help thinking of the hogs; they
> were so innocent, they came so very trustingly; and they were so very human
> in their protests—and so perfectly within their rights! ... It was like some
> horrible crime committed in a dungeon, all unseen and unheeded, buried
> out of sight and of memory.[56]

Sinclair's entrenched anthropocentrism yields him no room to label slaughter a "horrible crime" *simpliciter*—he must perceive the slaughtered animals as human-like, people being the presumed sole possessors of rights;
thus, the pig squeals sound "so very human" to his ear. As the passage
continues, Sinclair becomes even more overt in conflating nonhuman animals with humans, drawing from the ethical theory of human dignity and
the theological doctrine of human salvation as he struggles to comprehend

how he could be having a moral response to the mistreatment of a *nonhuman* animal:

> And each of them had an individuality of his own, a will of his own, a hope and a heart's desire; each was full of self-confidence, of self-importance, and a sense of dignity ... And now was one to believe that there was nowhere a god of hogs, to whom this hog-personality was precious, to whom these hog-squeals and agonies had a meaning? Who would take this hog into his arms and comfort him, reward him for his work well done, and show him the meaning of his sacrifice?[57]

When I first read this passage I took it to be sincere, if a little overblown. I was surprised to learn that Sinclair had later repudiated these sentiments, as he explains in his autobiography:

> For fifty-six years I have been ridiculed for a passage in *The Jungle* that deals with the moral claims of dying hogs—which passage was intended as hilarious farce. The New York *Evening Post* described it as "nauseous hog-wash"—and refused to publish my letter of explanation.[58]

This is a remarkable claim; if the passage was intended as "hilarious farce," then it was quite poorly written: it is not funny at all, and there is nothing in the book that sets up a reader to expect to find comic relief breaking out in the middle of Sinclair's unrelenting litany of industrial capitalism's crimes.

To me, the "hogs in heaven" passage reads not as hilarious farce but as a sort of nervous joke, an extended literary version of the sarcastic, half-smiling jibes we vegetarians are used to getting from certain meat eaters who, when confronted with the fact that moral opposition to animal slaughter exists, do not really know how to handle it. Sinclair is himself in that nervous position: as journalist he reports accurately on what he finds at the slaughterhouse; as novelist he cannot help but imagine his characters being horrified by the animal abuse. His inability to countenance a world beyond institutionalized animal exploitation, though, leaves him incapable of validating those reactions.

He leaves that tension unaddressed within *The Jungle*, but takes it up elsewhere. In *The Fasting Cure*, a book drawn from two articles he wrote for *Cosmopolitan* magazine in 1910 and 1911, Sinclair acknowledges that his research for *The Jungle* acquainted him with the "cruelty and filth" of the slaughterhouse, but denies that these are unavoidable features of the industry.[59] His rejection of morally based vegetarianism is expressed as follows:

> I had never taken stock in the arguments for vegetarianism upon the moral side. It has always seemed to me that human beings have a right to eat meat … What I ask to know about the question of meat-eating is the actual facts of its effect upon the human organism …; also, of course, its cheapness and convenience as an article of diet.[60]

Sinclair devotes several pages to making a case for the putative human "right to eat meat." His shallow arguments follow the usual specious lines: collective vegetarianism will lead to overpopulation of rabbits; animals exploit each other, so why can't we; domestic animals are only here to benefit us; meat production can be done humanely; vegetarianism involves cruelty to plants; and, "dumb animals" do not suffer as we do.[61] The last point, if analogously carried over to enslaved people of color, becomes exactly the argument Stowe was most concerned to counter, a final irony for the author whose most significant work was consciously modeled on *Uncle Tom's Cabin*.

CONCLUSION: TOWARD NEW NARRATIVES

In my undergraduate philosophy courses, I typically show videos during the unit on animal rights—footage of the inside of factory farms, slaughterhouses, animal research labs, and the like—to make the discussion more concrete, as well as to inform students about what's going on behind closed doors. I recall a lecture where after showing one particularly gruesome video, I opened discussion by asking how the class felt knowing that that kind of violence was behind the typical American bacon-and-egg breakfast. A quiet young man at the back of the room raised his hand, and, sincerely puzzled, asked, "What violence?" I had not been trying to ask a loaded question, or to use tendentious terminology, but had naively presumed that regardless of one's prior position on animal rights and vegetarianism, anyone seeing that video would agree that chickens crammed into cages and chained-up pigs getting their throats cut count as violence. That was a mistake. As I learned that day, two people can watch the exact same video while only one sees it as violent.

Several years prior to that, while a graduate student at the University of Pittsburgh, I made an appointment to talk with one of our professors, an intelligent and accomplished moral philosopher visiting from England. I wanted to be sure to sit down with him before his one-year appointment ended, to pick his brain on animal rights theory, as I was still sorting out the ideas for my dissertation. In his office we worked through several of the major ethical frameworks, discussing whether nonhuman animals would be accorded ethical protection according to their various terms. I recall extemporizing a utilitarian argument for vegetarianism, based on a comparison of the modest (or nonexistent) pleasure gained by one person over a lifetime of meat-eating against the lost future pleasures of the many chickens and other animals killed to furnish the diet. It seemed like the sort of rationalistic, quasi-mathematical argument that might appeal to this drily cerebral, very British philosopher. As I worried over the appropriate number of chickens to use in the utilitarian calculation, he waved me off. "It doesn't matter, Brian," he said. "It could be ten thousand, it still wouldn't make any difference. They're chickens. They just don't matter."

Who does matter? What counts as violence? How we answer these questions determines whether we are able to see issues regarding animals in moral terms. And, for many people, these answers are already given, prior to discussion and debate, operating so deeply they affect not just thoughts and beliefs but even perceptions themselves.

But, if the conviction that "chickens don't matter" does run that deep—conditioning the debate rather than resulting from it—that does not yet mean the conviction is innate. It still comes from somewhere. It originates, presumably, in the same way that most of our beliefs do, through what we see and hear around us, and most particularly, through what we hear people saying around us as we move through life. Our worldviews are shaped by what we tell each other, in word and deed, formally and informally, through the stories we pass on in conversation, in class, in court, from the pulpit, in news reports, through movies, television, and other media, and, of course, through our literary heritage.

The stories of the Western literary tradition are no friends to animals. They are functional, putting animals in their place, down at the bottom of the Great Chain of Being, that metaphysical abstraction through which human groups justify kicking and stomping those stationed below them. The four stories I examined in this chapter are not special in this regard—they were not chosen for analysis because I knew beforehand how each portrayed animals. I chose them as prominent narratives about four institutions of *human* life—child sacrifice, patriarchal marriage, slavery, wage labor—and asked the question, do they also have something to say about the place of animals within the nexus of intra-human domination? I have enough background in this kind of study to suspect that they would, but even so, the degree of correspondence between the four stories was impressive. Whether from ancient Greece, Elizabethan England, pre- or post-Civil War USA, each author used animals instrumentally in the same ways:

as background props, as a reservoir of stereotyped images for depicting humans, and, of most significance for our work, as paradigms of creatures exploitable by their very nature. The central stories of the Western tradition teach us, generation after generation, that animals exist to be taken by us.

This realization helps us answer the question posed by this volume: How do we move forward? In addition to our political, legal, economic, philosophical, and educational efforts to counter the institutions of animal exploitation, we must also be engaged *narratively*. The stories propagated in our culture serve to generate the anthropocentric worldview, a perspective by which a person can casually accept factory farming on the grounds that "they're just chickens," or watch a slaughterhouse video and see no violence. Our culture's stories must be addressed; they must be part of our strategic opposition to animal exploitation.

This is not an argument for censorship. Censorship is regressive, ineffective, and not even a real possibility until after the struggle has already been won, at which point it becomes a gratuitous violation of expressive freedom. Moreover, given the ubiquity of animals being presented instrumentally within our culture's stories, the question of what to ban quickly becomes a matter of what *not* to ban. I suppose Ursula K. LeGuin's *Buffalo Gals* might survive purging by an Animal Rights "Ministry of Public Enlightenment," along with a few other books and some movies. But we cannot live on *Bambi* and *Chicken Run* alone, can we?

Our example should not be Plato, who would have seen all poetry banned, or the Puritans, who actually did shut down the English theaters once they got control of the government. I would emulate Ellen McLaughlin, whose retelling of *Electra* includes that great, resonant line, "What happened to the horse?" or Andrei Serban, giving the people what they want, yet another production of *Taming of the Shrew* (and, indeed, one entirely true to Shakespeare's script), but with a little nudge for the audience at the end, as the actress playing Katherine walks off arm-in-arm, happily in love, not with the fellow who played Petruchio, but with one of the female cast members.

There is something to be said for retelling the old stories, as opposed to making up original material, given our limited capacity for taking in that which is truly novel, and our apparent preference for variations of the old, well-known tales. One of my own efforts in this direction is a retelling of the Akedah legend, the biblical story of the binding of Isaac.[62] This version of the Akedah includes all the text of the biblical account of Abraham's near-sacrifice of his son Isaac, as recounted in the book of

Genesis, Chap. 22, but with additional, interpolated lines I imagine to have been expunged over the years. The result is a "restored" Akedah in which the marginalized human characters—Abraham's wife and his son—regain their agency, with salutary results for the truly voiceless one in these ancient myths, the sacrificial animal.

The praxis of storytelling is *political*, notwithstanding that it eschews the *governmental* by involving neither the direct application of state force (the enforcement of laws), nor its indirect use (lobbying for statutory protection of animals, electing pro-animal candidates, etc.). The narrative approach is deeply political in the anarchistic sense of recognizing that radical social change is neither initiated nor consolidated by the state. Laws are meaningless without a general will of the people to comply and a general willingness of the executive and the judiciary to enforce. This is not to say that efforts to pass progressive legislation in advance of public opinion are otiose; indeed, legislative reform is itself, among other things, a form of narrative, as are other forms of governmental action such as court decisions and election campaigns, all of which can thus be useful as part of the work of telling new stories about justice and humanity.

We are the media through which culture is transmitted, filtering the great stream of stories that flows from one generation to the next. As each story enters our ears and passes through our lips, it is changed by that process, consciously or unconsciously, minutely or broadly, those changes themselves becoming part of our legacy to posterity. Yes, it is always possible, as we retell tales so that animals are no longer mere instruments but subjects of their own lives, that our changes will be misunderstood or rejected by the recipients: we have seen, after all, how Harriet Beecher Stowe's character of Uncle Tom has been stripped of his dignity, and how Upton Sinclair offered the world a detailed, 300-page indictment of capitalist exploitation, only to have the majority of readers react with, "Oh my God, there might be rat shit in my hot dog!"

People hear what they want to hear, so most of what we say will be rejected or misconstrued, most of the time. But, not always. Like-minded and open-minded individuals will hear our retellings, appreciate them, and pass them on, adding their own progressive modifications in the process. Bit by bit, a new Circle of Life zeitgeist will grow and spread, gradually, but inexorably, supplanting the outmoded, Great Chain of Being metaphysic, until one day we wake up, turn on our newsfeed, and find there a crime story from Reuters, just underneath the McDonald's banner touting their new all-vegan menu, and beside an item on the last Sea World park conceding the inevitable and closing its gates for good. We click on it, this little piece about a fatal shooting, a long-running property dispute ending in violence. It's a typical crime story, well-written, with all the right parties quoted, the district attorney, the defense lawyer, the witnesses. Nothing too remarkable here, except possibly for those of us who still remember the bad old days of anthropocentrism. We might notice something conspicuously absent from all those quotes, a term of hate, once casually applied to those deemed to fall outside the moral pale, but now absent, dropped at some point, evidently, from our lexicon of unreflective repression.

Nowhere in the story is the killer described as an animal. No one any longer even thinks to call him that.

Now he's just a man.

A man with a history of mental illness.

NOTES

1. See, for example, Alasdair Cochrane, *Animal Rights without Liberation: Applied Ethics and Human Obligations* (New York: Columbia University Press, 2012).
2. Joe McDonald, "Pennsylvania Man Gets Life for Rampage at Pocono Municipal Meeting," *Reuters*, May 29, 2015, accessed July 8, 2015, http://www.reuters.com/article/2015/05/29/us-usa-pennsylvania-shooting-idUSKBN0OE2LY20150529.
3. Cavalieri, this volume (p. 16).
4. Aeschylus, *Agamemnon* (458 BC), in *The Complete Aeschylus, Volume 1*, trans. Alan Shapiro (Oxford University Press, 2011), 52.
5. Euripides, *Iphigenia at Aulis* (408 BC), trans. Nicholas Rudall (Chicago: Ivan R. Dee, 1997), 48.
6. Colin Teevan, *IPH …* (London: Oberon, 1999), 64.
7. Ovid, *Metamorphoses* (8 AD), trans. Charles Martin (New York: W.W. Norton, 2004), 408.
8. Euripides, *Iphigenia at Aulis*, 66.
9. Walter Burkert, "Greek Tragedy and Sacrificial Ritual," *Greek, Roman, and Byzantine Studies* 7 (1966), 87–121.
10. Albert Henrichs, "Animal Sacrifice in Greek Tragedy: Ritual, Metaphor, Problematizations," in *Greek and Roman Animal Sacrifice: Ancient Victims, Modern Observers*, ed. Christopher Faraone and F.S. Naiden (Cambridge University Press, 2012), 181–182.
11. See, for example, Brian Luke, *Brutal: Manhood and the Exploitation of Animals* (Urbana: University of Illinois Press, 2007).

12. Ellen McLaughlin, *The Greek Plays* (New York: Theatre Communications Group, 2005), 21–22.

13. Ibid., 61.

14. Ibid., 54–55.

15. My unpublished study of the recent performance history (2010–2015) of 87 major North American classical theater companies found 1011 Shakespeare productions; the 38 canonical Shakespeare plays were thus produced, on average, 27 times each over this survey period. *Taming of the Shrew*, though, was produced 56 times, more than twice the average rate, behind only *Romeo and Juliet* with 69 productions and *A Midsummer Night's Dream* with 68, but surpassing standards of the repertoire such as *Macbeth* and *Hamlet* (55 and 50 productions, respectively), while eclipsing obscure Shakespeare works such as *Timon of Athens* (8 productions), *Coriolanus* (7), and *Henry VI, Part 3* (at the bottom of the pile, with just one lonely production, in 2011, at the American Shakespeare Center in Staunton, Virginia).

16. Justin Cord Hayes, *The Unexpected Evolution of Language* (Avon, MA: F+W Media, 2012), 195.

17. William Shakespeare, *The Taming of the Shrew*, ed. Barbara Mowat and Paul Werstine (New York: Simon & Schuster, 1992), 4.1.188–209 (references are to act, scene, and line numbers).

18. Ibid., 2.1.291.

19. Ibid., 2.1.267, 271.

20. Ibid., 3.2.235–238.

21. Jonathan Bate and Eric Rasmussen, ed., *William Shakespeare's The Taming of the Shrew: The RSC Shakespeare* (New York: The Modern Library, 2010), 145.

22. Shakespeare, *Taming of the Shrew*, Induction.1.118–122.

23. Ibid., Induction.2.41–45.

24. Ibid., Induction.1.19–25.

25. Ibid., 5.2.74–76.

26. Shirley Nelson Garner, "The Taming of the Shrew: Inside or Outside of the Joke?" in *The Taming of the Shrew: A Norton Critical Edition*, ed. Dympna Callaghan (New York: W.W. Norton & Company, 1988), 215.

27. Harriet Beecher Stowe, *Uncle Tom's Cabin: or, Life Among the Lowly* (New York: W.W. Norton, 1994), 283.

28. See, for example, Jennifer Holland's *Unlikely Heroes* (Workman Publishing Company, 2014), Rachel Buchholz's *True Love* (National Geographic, 2013), Stephanie LaLand's *Random Acts of Kindness by Animals* (Conari Press, 2008).

29. C. Vann Woodward (1955), *The Strange Career of Jim Crow* (Oxford University Press, 2002), 51.

30. Stowe, *Uncle Tom's Cabin*, 111.

31. Ibid., 341.

32. Ibid., 88.

33. Ibid., 80.

34. Ibid., 92.

35. Ibid., 288.

36. Ibid., 332.

37. Ibid., 300.

38. Ibid., 332.

39. Michelle Alexander, *The New Jim Crow: Mass Incarceration in the Age of Colorblindness* (New York: The New Press, 2010), 199–200.

40. Ibid., 113.

41. Ibid., 175.

42. Quoted in Eric Schlosser, "Foreword," in *The Jungle*, by Upton Sinclair (New York: Penguin Classics, 2006), xxvi.

43. Upton Sinclair, *The Jungle* (New York: W.W. Norton, 2003), ix.

44. Ibid., 210, 111, 292.

45. Ibid., 148, 160, 185, 290.

46. Ibid., 62, 67, 118, 130, 260.

47. Ibid., 155.

48. Ibid., 158.

49. Ibid., 154, 206, 259.

50. Ibid., 290.

51. Ibid., 103.

52. Ibid., 209.

53. Ibid., 215.

54. Ibid., 286–287.

55. Ibid., 34.

56. Ibid., 36.

57. Ibid., 37.

58. Upton Sinclair, *The Autobiography of Upton Sinclair* (New York: Harcourt, Brace & World, Inc., 1962), 164.

59. Upton Sinclair, *The Fasting Cure* (London: W. Heinemann, 1911), 145.

60. Ibid., 141, 146.

61. Ibid., 142–146.

62. See "The Lord Will Provide," posted February, 2016, at Brian-Luke.com.

Counter-Conduct and Truce

Dinesh Joseph Wadiwel

It is recorded that Michel Foucault did not read out his whole written lecture at the College De France on 1 March 1978.[1] Foucault concluded the lecture hurriedly by providing a quick gloss of his notes, and offering an apology to his audience: "There you are. Forgive me for having taken too long, and next time, this is a promise, we won't speak any more about pastors."[2] However, Foucault's lecture notes reveal that he had more to say; indeed, he excluded from his spoken lecture a fascinating contemplation of strategy and its relationship to truth:

> Rather than say that each class, group, or social force has its ideology that allows it to translate its aspirations into theory, aspirations and ideology from which corresponding institutional reorganizations are deduced, we should say: every transformation that modifies the relations of force between communities or groups, every conflict that confronts them or brings them into competition calls for the utilization of tactics which allows the modification

D.J. Wadiwel (✉)
School of Social and Political Science,
The University of Sidney, RCS Mills Build.,
NSW 2006, Australia
e-mail: dinesh.wadiwel@sydney.edu.au

© The Author(s) 2016
P. Cavalieri (ed.), *Philosophy and the Politics of Animal Liberation*,
DOI 10.1057/978-1-137-52120-0_7

187

of relations of power and the bringing into play of theoretical elements which morally justify and give a basis to these tactics in rationality.[3]

Here, political action does not merely require access to a "correct" version of truth that might circumscribe the appropriate tactics for change. Instead, every relation of power, every form of political context, requires a distinct appraisal of the political situation in question, generating different tactics and strategy, and with all of this, a theorisation of the prevailing forms of rationality, which might prompt and support the case for change. Here, the battleground for change is not merely a question of demanding the translation of ideals into practice; on the contrary, political change must deal with a shifting terrain, which is composed of both force relationships and evolving rationalities. It is not merely the challenge of political opinion or institutions we must contend with, but it is the challenge of truth itself.

Today, pro-animal advocates face an extraordinary challenge. On one hand, we confront massive systemic forms of violence directed towards almost all animals that humans have contact with, encompassing food systems, recreation, and scientific endeavour. This violence is so widespread and deeply embedded within institutions and practices that it almost seems impossible to imagine change: where might we begin? Simultaneously, we face a political challenge that relates to truth. While this systematically inflicted violence against animals is almost everywhere, almost nobody seems to care. Humans are either happy to be ignorant of this violence, or, alternatively, happily inflict this violence convinced of their own superiority, the supremacy of their own needs and pleasures, or confident of their prerogative over other creatures. Pro-animal advocates remain in an absolute minority with respect to their critique of the rationality that underpins the "animal industrial complex."[4] In this sense, the politics of animal liberation carries a strange affectation: pro-animal advocates are confronted and convinced by a horrible truth regarding human violence against animals; however, perhaps like the truth that Hamlet carried about King Claudius, this is a secret horror whose public reception is thwarted by its unimaginability.[5]

In this chapter, I offer an expanded discussion of the concepts of "counter-conduct" and "truce," as I had initially proposed in my book, *The War against Animals*.[6] An ongoing concern for me in thinking through these concepts is how pro-animal advocates might develop strategies that respond to different sites of contestation: inter-subjective, institutional, and epistemic (in a sense, this is a variation of the frame that Paola Cavalieri offers—this volume, Chap. 1—between material violence and symbolic violence[7]). As I have suggested earlier, and taking into consideration the Foucauldian perspective, the terrain of power cannot be disentangled from the terrain of truth; this is the same landscape in which we find rationality and politics intertwined. This chapter is divided into three sections. Firstly, I offer an outline of the main theoretical trajectories I offer in *The War against Animals*, which grounds my perspective on how we might think about the politics of animal liberation. Secondly, I offer an expanded elaboration of the concept of "counter-conduct," a concept I derive from Foucault's 1 March 1978 lecture I have described above. Here, I offer a more detailed explanation of the forms of counter-conduct that Foucault describes in his lecture, particularly with respect to forms of resistance against pastoral power in the Middle Ages. As I suggest in this chapter, it is my view that these modes of counter-conduct remain relevant to thinking through strategies, particularly tactics such as inter-subjective forms of conduct, including the pursuit of plant-based diets. Thirdly, in this chapter I provide an exploration of strategies for "truce." The thought experiment I offer here in the form of a proposal for a one-day truce in animal killing is partly modelled on an address made by Andrea Dworkin in 1983. In this context, I examine the labour general strike, its resonance with a potential animal liberation campaign for "one day without killing," and the potential alliance politics that might be involved in generating such an action. As I argue in this chapter, with reference to actions such as the 2015 veterinarian strike in Iceland (which led to "a lack of fresh meat in shops because pork and poultry abattoir shipments...[were]... halted"[8]), a proposal for a truce in the war against animals need not be an idealistic unachievable demand, but might instead be materially realisable.

War Against Animals

In many respects the main impulse behind *The War against Animals* was an assertion I have heard made by a variety of activists and scholars that our fundamental relations with animals represent a form of deep hostility and conflict. Animal advocates, for example, have frequently described human violence towards nonhuman life as "a war on animals"[9] (a powerful example of this is the recent documentary *The Ghosts in Our Machine*, where the main protagonist of the film, Jo-Anne McArthur, proclaims that her task of documenting the conditions faced by animals across the globe has transformed her into a "war photographer").[10] In *Eating Animals*, Jonathan Safran Foer has framed industrialised meat production as a war: "We have waged war, or rather let a war be waged, against all of the animals we eat. This war is new and has a name: factory farming."[11] While in his last lectures, philosopher Jacques Derrida identified the Western philosophical tradition as enabling a war against animals, remarking that the "Cartesianism belongs, beneath its mechanicist indifference, to the Judeo-Christiano-Islamic tradition of a war against the animal, of a sacrificial war that is as old as Genesis."[12] These perspectives are in many ways cognizant of the immense material violence generated by our attitude towards other creatures, which represents an intense, ever-present, but frequently unacknowledged, relationship of hostility. Indeed, given the magnitude and near-universal dimension of human violence towards animals, it is not unreasonable to suggest that our mainstay relationship with animals is making them suffer and die for our own purposes. From this perspective, perhaps the intuition that "we are at war with animals" does not appear to be outrageous.

But this intuition felt by many pro-animal scholars and activists (that we are at war with animals) might also be supported by theoretical perspectives on war and its relationship to peace. It is, perhaps, interesting to note that "war" has been a continuing concern within modern political theory. Thomas Hobbes, for example, defines the unifying force of sovereign power by distinguishing it from a chaotic state of nature that is described as a war of "all against all" (*bellum omnium contra omnes*).[13] In a different context, Niccolò Machiavelli considers the civil political sphere as a site for the continuation of war-like tactics, particularly in relation to weighing the judicial application of force against the manipulation of consent. Here, tactics in war and tactics in peace are considered interchangeably; arguably, this is why so much of *The Prince and the Discourses* is devoted to evaluating the use of violent as opposed to the use of non-violent means,[14] or trust and loyalty against the utility of ruling by fear.[15] War and peace sit at the centre of important theories of sovereignty, such as that advanced by Carl Schmitt[16]; this has proved to be influential for theorists such as Jacques Derrida, who reads an inter-relationship between war, politics, and friendship.[17] The distinction between war and peace is a theme in postcolonial and critical race scholarship: Achille Mbembe, for example, locates Western sovereignty within a globalised context of colonisation and racialised terror,[18] and Aileen Moreton-Robertson suggests that race war is the key to understanding the evolution of rights discourse and the reframing of the question of indigenous sovereignty.[19]

The particular trajectory on war that I found most compelling for my own analysis was derived from the work of Michel Foucault. In his 1975–1976 series of lectures, published in English under the title *Society Must Be Defended*,[20] Michel Foucault devoted a great deal of space to thinking about the relationship between war and politics, particularly the ongoing relationship within political theory and the theorisation of conflict. It is in these lectures that Foucault outlines a model for understanding sovereignty and its relationship to population as an essentially antagonistic relation, a template that would underpin his development of the concept of "biopolitics." From this flows an understanding of politics within societies as war under the guise of peace. Here, Foucault is essentially offering a critique of traditional models of liberal society founded on agreement or compact:

> War is the motor behind institutions and order. In the smallest of its cogs, peace is waging a secret war. To put it another way, we have to interpret the war that is going on beneath peace; peace itself is a coded war. We are therefore at war with one another; a battlefront runs through the whole of society, continuously and permanently, and it is this battlefront that puts us all on one side or the other. There is no such thing as a neutral subject. We are all inevitably someone's adversary.[21]

In this context, Foucault provides an understanding of sovereignty that follows from the annihilatory violence of war. Foucault notes that political sovereignty is founded upon a material and real set of "victories" in war-like contexts (e.g. when one nation invades another). Here, the outcome of war is not peace but deeply entrenched relations of power, with concessions granted to those who are dominated only in the face of the threat of death. Foucault observes:

> The vanquished are at the disposal of the victors. In other words the victors can kill them. If they kill them, the problem obviously goes away: the Sovereignty of the State disappears simply because the individuals who make up that State are dead. But what happens if the victors spare the lives of the vanquished? If they spare their lives, or if the defeated are granted the temporary privilege of life...[they]...agree to work for and obey the others, to surrender their land to the victors, to pay them taxes. It is therefore not the defeat that leads to the brutal establishment of a society based upon domination, slavery, and servitude;...It is fear, the renunciation of fear, and the renunciation of the risk of death. The will to prefer life to death; that is what founds sovereignty.[22]

In this view (and against liberal political theory approaches), there is no social compact that founds sovereignty. Foucault observes, in language that seems to mirror the terse text of Hobbes *Leviathan*, that battle and violence form the basis of political orders: "War obviously presided over the birth of States: right, peace, and laws were born in the mud of battles…The law is not born of nature, and it was not born near the fountains that the first shepherds frequented: the law is born of real battles, victories, massacres, and conquests…the law was born in burning towns and ravaged fields."[23] Here, instead of accord and agreement founding social relations, Foucault offers us a perspective on sovereignty as established on the basis of conflict and violent inequality. Sovereignty seeks to guarantee continuing forms of victory in "peace" over those who have been dominated through this conflict. In this view, Foucault provides a way to think about how sovereignty and the apparent "peace" it ushers, epistemologically shield deeply structured violence to make it appear as if it were not present. War is conducted under the guise of peace: "politics is war pursued by other means."[24]

This perspective advanced by Foucault is potentially useful for understanding how it is that liberal societies, apparently founded upon principles of equality, freedom, and democratic peaceability, should shield and tolerate (indeed perpetuate) systemic forms of violence against particular population groups.[25] However, in addition to providing a critique of liberal perspectives on the genealogy of State power, Foucault also provides a remarkably useful template for thinking about our relationships with animals. Here, Foucault's conceptualisation of war and sovereignty offers us a method for analysing the practice of mass orchestrated violence where this is seamlessly integrated within civil political spaces in an imperceptible way. As such, in my view, this helps to make sense of both our structural commitment to doing violence towards animals for our own ends and our simultaneous disavowal of this violence (it is so commonplace, nobody seems to notice this violence, or indeed care). Importantly, Foucault provides a rationale for this organisation of violence: the inherent resistance of human institutions to change, even in the face of seemingly rational and fair argument, is reflective of the massive human investment in the continuation of violence and the costs (for humans) associated with change (in the loss of privileges that have been won through violence).

This framework for understanding our political relationships with animals is potentially productive for reappraising some different theoretical conceptions, including biopolitics, property and commodity relations, and sovereignty. By speaking of biopolitics, I am referring to the concept developed by Foucault in his lectures and in the final chapter of his important *History of Sexuality Volume One*. Foucault described biopolitics as a shift away from a directly coercive model of power focused on the acquisition of territory and resources (as perhaps best conceptualised by the violent sovereign power of the king), towards a rationality that attends to populations and life. Foucault enigmatically marks this shift by arguing that "the ancient right to *take* life or *let* live was replaced by a power to *foster* life or *disallow* it to the point of death."[26] Here, Foucault suggests that political discourse is increasingly directed towards, and marked by, a concern for the nature of *life*, its vicissitudes, its requirements, its essence: "For millennia, man [*sic*] remained what he was for Aristotle: a living animal, with the additional capacity for a political existence; modern man is an animal whose politics places his existence as a living being in question."[27] This understanding of biopolitics has much to say about animals; indeed, I would argue that our hostility towards animals is distinctly *biopolitical* in character, in a somewhat paradigmatic way. There are at least two ways that this can be understood. Firstly, the political sphere itself is constructed through a continual and violent separation between human and animal. Thus, the way in which we construct politics, the political subject, and consider membership of the political sphere, requires a systematic, continuing, and violent division between human and animal.[28] More recent developments in Foucault's understanding of biopolitics, such as the famous account Giorgio Agamben offers in *Homo Sacer*,[29] perhaps only underline this reality: Agamben states in *The Open* "the decisive political conflict, which governs every other conflict, is that between the animality and the humanity of man. That is to say, in its origin Western politics is also biopolitics."[30] Secondly, it seems impossible to escape the sense in which the mechanisms of human violence towards animals appear to be distinctly biopolitical in character. For example, in the case of large-scale animal-based food production, it seems clear that industrialised slaughter (making death) is interdependent on industrialised reproduction (making life). This means that as the scale of industrialised killing increases (almost exponentially it would seem), then the scale of industrialised breeding also increases. The more animals we kill for food, the more we need to breed; the faster we kill animals, the faster we need to breed them.[31] In this context, we might further extend Foucault's understanding of the rationality that develops for the management of population—"governmentality"—in connection with the genealogy of human domination of animals: governmentality is a useful way to think about human management of nonhuman populations towards life and death systems of value creation.[32] Later in the text, I will further explore governmentality and "pastoral power."

Another trajectory that is relevant for me is the conceptualisation of animals as property and commodities and the implications of this for thinking about animals within political relations. Of course, within animal rights theory, property rights in animals has been an area of theorisation, notably in the work of Gary Francione, who has provided a useful analysis of the way in which welfare considerations spring from a human use value for animals as property.[33] I offer an extension of this perspective, drawing together both the biopolitical perspective (particularly that of Roberto Esposito[34]) and the view of sovereignty that I have described earlier. Here, I believe we can extend Francione's analysis through a close reading of John Locke, who arguably provides the template for a contemporary conception of property within political theory and law.[35] I argue that property represents the everyday form of appropriation by which humans claim dominion over animals. Importantly, and this is the connection to contemporary biopolitics, it is within the conceptualisation of animals as property that humans find a form of sovereign immunity from a nonhuman claim on the commons of the political sphere (in a crude sense, human "rights" are founded upon the continuing denial of the rights of nonhumans). In other words, human politics itself uses property rights in animals to insure itself against the entry of animals to the political sphere. This appropriation gives way to a system of values including a market system that is premised upon, and secured by, animal domination. In this sense, violence against animals, the processes of violence that are part of the "subsumption" of animals into markets as commodities of exchange and bearers of congealed values, is an important preliminary step; war is required in order for animal bodies to be put to market as part of a process of "original accumulation." As Karl Marx reminds us, surplus is the driving principle for capitalist economic exchange, and conflict is generated for control over the surplus extracted. However, in the case of animals, this surplus is generated through a primary exploitation that precedes human labour and exchange, through a violent conversion of the nonhuman animal into a simple resource for consumption and conversion within the production process. When Marx comments in *Capital* that "No boots can be made without leather,"[36] we must assume that the processes of violence and subordination that lead to the production of leather pre-exist—indeed discursively and "naturally" pre-empt—the human labour production process. This is secret war that relies on forgetting to harmonise social relations and generate values.[37]

Sovereignty is a continuing theme in the perspective I offer in *The War against Animals*. Indeed, the book offers a reconceptualisation of political sovereignty as a specific systemic form of human domination of nonhuman life. From my point of view, the term *sovereignty* usually implies a mandated system of authority, rule, or control that provides a nodal point for the organisation of political and social relations. This view is reflective of contemporary understandings of sovereignty within political theory, which are, in many ways, shaped by the Western—distinctly Hobbesian[38]—conception, namely, the perceived right of a delegated authority to make a unifying claim to life and territory within a particular jurisdiction or domain, and to both organise and exercise legitimised violence. Within recent animal rights theory, there have been attempts to recuperate sovereignty as a concept for imagining animal self-determination, including in Robert E. Goodin, Carole Pateman, and Roy Pateman's proposal[39] for simian sovereignty and, more recently, in Sue Donaldson and Will Kymlicka's proposal for animal citizenship, denizenship, and sovereignty.[40] Distinct from these perspectives, I have offered a different way of imagining sovereignty not simply as a juridical apparatus for humans to exert rule over territory or political subjects, but as potentially a mode of human domination of animal life. As I have argued earlier, and drawing explicitly from Foucault, I have argued that our war against animals might be conceptualised as a mode of sovereignty that internalises a continuing combat for the benefit of ongoing human utility; here, intense forms of violence against animals are stabilised in such a way to appear peaceable. Animals might also be considered as exercising sovereignty. If we understand human sovereignty over animals as discursively erasing the possibility of animal sovereignties, then it is possible to conceptualise animal sovereignties as a strategic response. However, to understand "animal sovereignties," I have moved away from a conception of sovereignty found in liberal political theory and, instead, examined Jacques Derrida's treatment of sovereignty in his final *The Beast and The Sovereign* lectures.[41] I believe Derrida provides a quite distinct perspective in these lectures, particularly in the argument that sovereignty is essentially groundless: sovereignty lacks content, and is not founded upon any essential capability in those who make claim for this sort of legitimation.[42] On the contrary, sovereignty operates as a kind of right to "stupidity." Humans claim sovereignty over animals because we have prevailed over animals using force, and not because we actually have any "superiority" in capability. On one hand, this argument provides a useful basis for describing the somewhat nonsensical claim of superiority that humans command over nonhuman animals. (What better way to describe a claimed superiority by humans over animals [whether based on intelligence, reason, communication, vocalisation, or politics] that has no consistent or verifiable scientific or philosophical basis?) On the other hand, if there is no consistent justiciable basis on which humans make a claim of sovereignty,[43] it becomes imaginable that animals might exert their own sovereignty; indeed, animals might be said to always present their own form of countervailing force, their will for self-preservation, and their own sovereign power.

And here is one final extension of my view of sovereignty and its implications for imagining animal sovereignty, namely, in the concept of resistance. There has been some interesting scholarly work within animal studies, which has dealt with the question of animal resistance and its relationship to power. A pioneering scholar in this area is Jason Hribal, who documents through historical case studies examples of animals breaking free from human control—breaking down fences, escaping abattoirs, tussling with human controllers, and maiming those who stand in their way.[44] However, I offer a different description of resistance, influenced by Marxist autonomist or "operaist" conceptions,[45] a perspective I have since expanded upon.[46] In my view, our war against animals is not one-sided, but represents a tussle for power between humans and nonhumans. This view highlights that power is never one-directional; on the contrary, power describes a series of complex relationships, potentially comprising the interaction between force and resistance. According to this perspective on resistance, animals do not give themselves up to us; on the contrary, humans must continually develop techniques, practices, technologies, and knowledges to counter the resistance of animals to "bend them towards our wills."[47] One example I offer of this is the curved corral, which has been developed as a welfare measure and is increasingly found within industrialised slaughter houses.[48] The rationale for the introduction of curves into the chutes that led animals to slaughter was to minimise the possibility of animals responding to the chute by balking and backing up.[49] While the argument for introducing curves into the races that led animals to slaughter might satisfy a demand for improved welfare (and arguably such designs do tangibly reduce suffering), the curved corrals can also be understood as a means of containing and dealing with animal resistances. The curves have been designed to counter animal insubordination to the logics of human utilisation; the curves are there to subdue resistance and enable the effective and smooth process of slaughter, maximising human utilisation (and profit) value. Here, in my perspective, animal resistance generates technological change; innovation in the animal industrial complex is founded upon continuing attempts by humans to counter animal insubordination to rule. While this presents a somewhat foreboding picture of the complete and encompassing power of human sovereignty, this perspective simultaneously allows us to recognise the possibility of animals as political agents of resistance and tells us something about animal sovereignties. It is because animals contest our own claim of sovereignty—a claim we arrogantly assume is impervious to challenge—that animals simultaneously declare their own sovereignty. In this sense, Foucault's phrase "the will to prefer life to death; that is what founds sovereignty" described earlier, although intended to describe the life and death terms by which political submission occurs within a political system, is, in fact, perfectly reversible. It is an absurdity to assume that animals prefer to die at our hands rather than live; indeed, it is the bare fact that animals assert their will to live in the face of our violence—chickens struggle against human operators before being thrust into hangars, cows hesitate before being prodded to enter kill chutes, and so forth—that suggests the existence of a sovereignty of animals, which contests with our own. In this regard, animal sovereignty need not be necessarily conceptualised as an end state that represents a juridical or territorial formation recognised within a global order (such as the nation state); on the contrary, animal sovereignty might simply be understood as a form of self-determining force, which human domination attempts to suppress within relations of power. All wars are life and death contests for self-determination; our war against animals is no different, and represents an active contestation of animal sovereignties, which we (humans) seek to overcome for our own benefit.

COUNTER-CONDUCT

One of the challenges to political action—which I discussed briefly at the introduction of this chapter—is that the war against animals is not merely inter-subjective (i.e. related to individual choices, action, or ethics) nor just institutional (i.e. related to the operation of large-scale industries, bureaucracies, political apparatuses, or corporations), but that also operates on an epistemic level. In this sense, while it is important to pay attention to inter-subjective violence and institutional violence, there is another task in paying attention to the operation of "epistemic violence." In understanding this concept of epistemic violence, I draw explicitly on Gayatri Chakravorty Spivak and the conceptualisation offered in her essay "Can the Subaltern Speak?"[50] Spivak describes "epistemic violence" as a way to understand the capacity of systems of truth to allow some to speak (and be heard) while silencing others, and render visible and invisible particular forms of rationality and possibility.[51] To explain this, Spivak offers the case study of the ritual of widow burning in India, *sati*, the practice that was subject to legal regulation by the British as part of their colonising mission in India, and then subject to response from Indian traditionalists claiming the practice as a "custom."[52] Within this example, Spivak draws attention to the way in which a system of truth shaped the narratives in such a way as to silence the voices of Indian women:

> The Hindu widow ascends the pyre of the dead husband and immolates her-self upon it. This is widow sacrifice. (The conventional transcription of the Sanskrit word for the widow would be *sati*. The early colonial British tran-scribed it *suttee*). The rite was not practiced universally and was not caste- or class-fixed. The abolition of this rite by the British has been generally understood as a case of 'White men saving brown women from brown men.' White women—from the nineteenth-century British Missionary Registers to Mary Daly—have not produced an alternative understanding. Against this is the Indian nativist argument, a parody of nostalgia for lost origins: 'The women actually wanted to die.'[53]

I am drawn to this citation from Spivak as it, I believe, offers some useful tools for understanding the political challenge of anthropocentrism, and the problem of how knowledge and violence render their subject. Systems of knowledge can work to effectively silence by failing to provide the sub-ject (or in Spivak's rendering "the sub-altern") the means to speak. Of course, in a direct sense, this is the goal of anthropocentrism: to render animals and their world only through the lens of the human.[54] Tom Tyler names this movement "epistemological anthropocentrism," a knowledge assumption that the human comes first and crafts the world through, and only through, human experience: "this is the epistemological claim that all knowledge will inevitably be determined by the human nature of the knower and that any attempt to explain experience, understanding, or knowledge of the world, of Being, of others—must inevitably start from a human perspective."[55] In sense, a continuing feature of the war against animals is the way in which knowledge systems render animals as always lesser, dumber, eatable, and killable. This epistemic violence is difficult to confront in so far as it is inherent to available knowledge systems and underpinned by the arrogance of human sovereignty; the possibility that animals might mean something beyond human representation, that animals may think and feel, that animals may be worth more than the use values we apply to them, appears to evade our knowledge systems.

This combination of the inter-subjective, institutional, and epistemic dimensions of violence that I have used to conceptualise our relations with animals forecloses the possibility of describing our relationships with animals as simply comprising a top-down system of domination. On the contrary, we are dealing with a power form comprising relationships, which envelop humans and animals, and shape a range of actions, from inter-subjective (acts of killing, experimentation, and pet keeping) to institutional (the development of large-scale industrialised apparatuses for breeding, containment, and killing; the development of large-scale supply and value chains within economies to price and distribute animals and animal parts as commodities) to epistemic (the continuing development of knowledge systems that reaffirm the superiority of human life, contest and debate the terms for thinking about ethics and politics, and reproduce technical knowledges of welfare practices within slaughter and experimentation industries). Political resistance to all of these levels of relationality—at least simultaneously—remains difficult to imagine. Certainly, traditional models of politics, where, for example, political subjects "take a stand" against known political opponents, do not necessarily fit models of action that might be effective in the war against animals. For example, when a young teenager announces at the dinner table that she will no longer eat animal flesh, who exactly is the political target of this decision? Is the action purely an ascetic gesture, designed as a form of politicised self-sacrifice or a personal challenge? Does the teenager simply prefer the taste of vegetables? Is the act levelled at her family, perhaps as an act of rebellion (to be rendered by her exasperated family as a "phase she is going through"), or is the political target of her action the vendors who might supply food at the dinner table? Is the act aimed at a large-scale producer or factory farm? Does the teenager exercise that much valorised "consumer sovereignty" in order to influence change? On the contrary, is it purely an ethical decision in a Kantian sense: "Killing animals is wrong and I do not want to have any part of it?" Does the teenager's decision seek to respond to an epistemological order that renders animals and their deaths as unimportant? Does this political action seek to generate a new set of truths about how it may be possible to live? I raise all of this to point out that the politics of resistance with respect to the "animal industrial complex" arguably operates within different fields of engagement from other forms of political activity, such as attending a strike or protest, or exercising a vote. In a sense, pro-animal political engagement can involve confrontation with inter-subjective spheres (the self and one's appetites/ desires, friends, family), institutions (corporations, political parties) and with truth itself (at least, the prevailing truth that animals do not matter).

In attempting to explore a mode of political engagement that seeks to address these different levels of inter-subjective, institutional, and epistemic violence, I have been attracted to Foucault's remarkable 1 March 1978 lecture, which I referred to in the introduction to this chapter. In my view, the exploration of *counter-conduct* in the lecture marks a significant shift in how Foucault understands the problem of resistance (and is one of Foucault's many subsequent reformulations of the problem of resistance and its relationship to subjectivity[56]). Relevant to my discussion here, the lecture offers some remarkable tools for understanding contemporary practices of resistance against the animal industrial complex, including the practice of veganism.

The context for the lecture is Foucault's shift of attention from disciplinary power (as described in *Discipline and Punish*[57]) and sexuality and normalisation (as summarised explicitly in the *History of Sexuality Volume One*[58]) towards an interest in the problem of government and its relationship to the conduct of the self. The 1977–1978 lectures, published in English under the title *Security, Territory, Population* began a long elucidation of the concept of "Governmentality," that would occupy Foucault for several years, and shape many of the philosopher's later concerns in relation to ethics and its relationship to liberalism. According to Foucault, and in line with the explanation of biopolitics I have provided earlier in the text, governmental rationalities mark a significant shift in the operations of State power, moving away from the maintenance of territory (as the mere physical possession of the sovereign) as a goal of power towards biopolitical intervention at the level of population as an end in itself. In contrast to a view that sovereignty is concerned merely with the maintenance of its own means of power, government has as its purpose not the act of government itself, but the welfare of the biological population, the improvement of its condition, and the increase of its wealth, longevity, health, and broad "well-being." Governance involves a strategy which Foucault suggests is "economic" in nature.[59] Here, rather than exerting force through spectacular bouts of violence that aim to inscribe and win power, governing is a rationality of management that seeks, by discreet interventions, to order the behaviour of populations in a relational field. Governmentality, thus, represents what has been described as "the conduct of conduct."[60] In this sense, Foucault will suggest that governmentality has a distinctly "pastoral" element: perhaps, best exemplified in the metaphorical relationship between a shepherd and livestock.[61] Government attempts to organise a flock, to develop strategies to manage the individual components within a multiplicity. The genealogical trail, which leads to the development of governmentality that Foucault identifies at some length in the 1977–1978 lectures, is traced to the tradition of the Christian church; indeed, Foucault's interest in "pastoral power" in the lectures is related to his view that this form of power is a precursor to governmentality.[62]

This is the context for the 1 March 1978 lecture, where Foucault turns his attention to the history of movements of "resistance" against orders of conduct within Europe in the Middle Ages. Here, Foucault ponders the specific forms of resistance that might have sought to insubordinate and challenge the reach of pastoral power:

> What is at issue? If it is true that the pastorate is a highly specific form of power with the object of conducting men—I mean, that takes as its instrument the methods that allow one to direct them (*les conduire*), and as its target the way in which they conduct themselves, the way in which they behave—if the objective of the pastorate is men's conduct, I think equally specific movements of resistance and insubordination appeared in correlation with this that could be called specific revolts of conduct...They are movements whose objective is a different form of conduct, that is to say: wanting to be conducted differently, by other leaders (*conducteurs*) and other shepherds, towards other objectives and forms of salvation, and through other procedures and methods. They are movements that also seek, possibly at any rate, to escape direction by others and to define the way for each to conduct himself. In other words, I would like to know whether the specificity of refusal, revolts, and forms of resistance of conduct corresponded to the historical singularity of the pastorate. Just as there have been forms of resistance to power as the exercise of political sovereignty, and just as there have been other, equally intentional forms of resistance or refusal that were directed at power in the form of economic exploitation, have there not been forms of resistance to power as conducting?[63]

In trying to develop this notion of "resistance to power as conducting," Foucault explores a range of alternative terms to try and encapsulate the specific movements,[64] finally settling on the formulation "counter-conduct." Foucault's interest in the phrase counter-conduct is the capacity to describe a political activity that works against multiple relations of power, that need not be individualising, and may lack the intentionality of a "deliberative" political action: "by using the word counter-conduct, and so without having to give a sacred status to this or that person as a dissident, we can no doubt analyse the components in the way in which someone actually acts in the very general field of politics or in the very general field of power relations; it makes it possible to pick out the dimension or component of counter-conduct that may well be found in fact in delinquents, mad people, and patients."[65] As I shall discuss later in the text, this perspective on resistance offers a way to think about how political action might address manifold and enveloping forms of power that cut across inter-subjective, institutional, and epistemic levels, and, as such, is useful for thinking about forms of resistance to the animal industrial complex, and the modes of human conduct inherent to this system of violence.

Foucault names five different modalities of counter-conduct in the lectures. *Firstly*, Foucault describes the development of different forms of ascetic practice in the Middle Ages (e.g. self-sacrifice and personal disciplines and prohibitions) as a form of counter-conduct against pastoral power. Admitting that it may appear odd to suggest that asceticism posed a threat to organised Christianity, Foucault points out that in his view the church worked to universalise teaching and discipline, and centralise authority within communities in such a way that aimed to prevent individual forms of ascetic practice. Here, individual ascetic practices contested the enveloping power and authority of the church:

> The organization of monasteries with obligatory communal life; the organization of a hierarchy around the abbot and subordinates who relay his power; the appearance of a communal and hierarchized life according to a rule imposed in the same way on everyone, or anyway on each category of monks in a specific way, but on all the members of that category, according to whether they are novices or elders; the existence of the superior's absolute, unchallenged authority with the rule of unquestioning obedience; the assertion that real renunciation is not renunciation of one's body or flesh, but essentially renunciation of one's will, in other words, the fact that the supreme sacrifice demanded of the monk in this form of spirituality is essentially obedience, all clearly show that what was at stake was limiting anything that could be boundless in asceticism, or at any rate everything incompatible with the organization of power.[66]

Here, ascetic practices, models of relationally with oneself, are designed to establish relationship between the subject and God that is not mediated by the church, effectively challenging the rationality of pastoral organisation inherent to the prevailing authority of the Church. Thus, according to Foucault, those who pursued asceticism developed their own modes of self-discipline as part of their spiritual practice (e.g. dietary or prayer regimes), effectively side-stepping the authority and knowledge systems inherent to pastoral power.

Secondly, Foucault observes that counter-conduct could be operationalised as a collective or communal form of resistance. While asceticism generated forms of individual counter-conduct, there were communal counter-conduct movements in the Middle Ages, found in the development of intentional communities and secret sects. As Foucault observes, the ongoing doctrinal tensions within the Church in the Middle Ages— such as the question of whether the Roman Catholic Church did indeed possess moral authority—themselves produced experimental communities, which attempted to resolve these tensions. As Foucault notes, this led to the development of sects that experimented with different modalities of organisation, including in developing new forms equality:

> there may be the principle of absolute equality between all members of the community, either in a religious form, that is to say each is a pastor, a priest, or a shepherd, which is to say nobody is, ... [or in the] ... strict economic ... [form]... that you find in the Taborites, in which there is no personal possession of goods and anything acquired is acquired by the community, with an egalitarian division or a communal utilization of wealth.[67]

An extension of this principle is the disavowal of obedience to the Church in favour of obedience to those comprising one's community: "there are, for example, relationships of reciprocal obedience. In the Oberland Friends of God there were rules, or oaths rather, pledges of reciprocal obedience of one individual to another. Thus Rulman Merswin and the anonymous Oberland Friend of God made a pact of reciprocal obedience for 28 years. For 28 years each agreed that they would obey the other's orders as if he were God Himself."[68] Here, just as in ascetic practices, counter-conduct strategies seek to side-step the authority of pastoral modes of organisation, and in the case of these communal practices, counter-conduct relies on autonomous forms of community and reciprocity.

Thirdly, Foucault observes that counter-conduct to pastoral power can take the form of mysticism, where the believer resists the hierarchical deployment of truths from within the church in favour of direct and unmediated forms of knowledge of God:

> Basically, pastoral power developed a system of truth that, as you know, went from teaching to examination of the individual; a truth conveyed as

dogma to all the faithful, and a truth extracted from each of them as a secret discovered in the depths of the soul. Mysticism is a completely different system. In the first place, it has a completely different game of visibility. The soul is not offered to the other for examination, through a system of confessions (*aveux*). In mysticism the soul sees itself. It sees itself in God and it sees God in itself.[69]

Here, counter-conduct is associated with resisting truths, and with them the passing down of truths through established orders, often by drawing on personal experiences and revelations, and "subjective" processing of knowledge: "The pastorate was the channel between the faithful and God. In mysticism there is an immediate communication that may take the form of a dialogue between God and the soul."[70] Connected with this, Foucault observes a *fourth* trajectory of counter-conduct, in the return to Scripture. He remarks that the pastorate drew up a privileged relation between the pastor and sacred written texts, where the Church relayed the interpretation of Scripture to the believer. Resistance to these processes in counter-conduct movements saw an insistent return to the text, to providing new translations and interpretations of the text: "In the movements of counter-conduct that develop throughout the Middle Ages, it is precisely the return to the texts, to Scripture, that is used against and to short-circuit, as it were, the pastorate."[71] Finally, Foucault observes that the development of eschatological beliefs comprise a *fifth* form of counter-conduct. Here, the imagination of pastoral power itself as a transitional stage towards a higher stage of human progress becomes a mechanism for disempowering (or time-limiting) the reach of pastoral power: "After all, the other way of disqualifying the pastor's role is to claim that the times are fulfilled or in the process of being fulfilled, and that God will return or is returning to gather his flock."[72]

In my view, Foucault's exploration of counter-conduct is immensely valuable for considering how it is that we might oppose the war on animals as a system that not only kills animals and makes them suffer, but orders human conduct across inter-subjective, institutional, and epistemic realms. Part of my reasoning here is that pro-animal advocates do not simply face a form of oppression where there is a single source of domination or site of contestation. Ending the war on animals does not merely involve confronting human sovereignty over animals, as if this were a matter of storming the barricades at the winter palace; nor does it mean ending an economic system of power, such as, for example, legislating against animals as property. Further, ending the war against animals does

not merely involve a contest of ideas, whether by challenging individuals to change their minds through rational discourse, or through shifting ideologies and challenging cultural hegemonies. On the contrary, in
confronting the war against animals, we are dealing with a set of violences
that are profoundly and deeply embedded into almost every conceivable
facet of human organisation, life, and knowledge in a way that exceeds
and incorporates sovereignties, economies, and information. The systemic
domination of animals is an encompassing reality that shapes who we are,
how we relate to others, and what we know about ourselves. In other
words—and this is my attraction to Foucault's understanding of counter-
conduct movements—resistance must seek to confront conduct: how we
are governed, how we govern ourselves, what we know about ourselves,
and what we know about others. Just as resistance to the totality of the
Christian church in the Middle Ages must have involved manifold contestations in every locale of pastoral power– the Church's engagement
with State power, the role of the Church in organising hierarchies, the
role of the Church in knowledges, the role of the Church in determining the appropriate relationship between the self and God—pro-animal
advocates currently face an extraordinary proliferation of potential sites of
contestation, which comprise all facets of an enveloping system of human
domination of animals, of which there are inter-subjective, institutional,
and epistemic dimensions of struggle.

One prominent site of contestation pertains to individual practices
within the war against animals, particularly decisions taken by individuals
to withdraw themselves from complicity with forms of violence against
animals. One example of this is the practice of veganism: where individuals exempt themselves from the violence of the war on animals by
refraining from eating and preparing animal-based foods, and refusing
to wear or consume animal-based products. In some respects, decisions
by individuals to pursue plant-based diets, or refrain from wearing animal-based textiles, appear as strategies of "desertion" from war. Here,
resistance involves attempting to exempt oneself from prevailing and
commanding disciplines, practices, and norms. Indeed, this sort of political activity of "desertion" is directly discussed by Foucault as an example
of counter-conduct:

> Desertion was an absolutely ordinary practice in all the armies of the sev-
> enteenth and eighteenth centuries. But when waging war became not just
> a profession or even a general law, but an ethic and the behaviour of every
> good citizen of a country, when being a soldier was a form of political and

moral conduct, a sacrifice, a devotion to the common cause and common salvation directed by a public conscience and public authority within the framework of a tight discipline; when being a soldier was therefore no longer just a destiny or a profession but a form of conduct, then, in addition to the old desertion-offence, you see a different form of desertion that I will call desertion-insubordination. Refusing to be a soldier and to spend some time in this profession and activity, refusing to bear arms, appears as a form of conduct or as a moral counter-conduct, as a refusal of civic education, of society's values, a refusal of a certain obligatory relationship to the nation and the nation's salvation, of the actual political system of the nation, and as a refusal of the relationship to the death of others and of oneself.[73]

In many ways, veganism, at least in its practice within Western societies,[74] carries with it some of the characteristics of desertion and appears as an example of both a current practice of counter-conduct, and one of the possible avenues for critical refinement as a strategy to resist the war on animals. Vegan practice is distinctive in so far as it is not merely reflective of a set of political beliefs or ideology, nor challenges a single agent or system of authority, but works across different levels of conduct. Many of these modes of conduct appear to resonate strongly with the trajectories of counter-conduct described earlier in the text. Vegan practitioners typically modify their ways of living and consumption substantially against prevailing norms, frequently facing forms of social isolation due to dietary and other choices. In many respects, this reflects the forms of resistant asceticism that Foucault describes as a central counter-conduct strategy during the Middle Ages. Here, how one "conducts oneself" in the face of prevailing and near-universal disciplinary norms and practices—including prevailing norms that reinforce the normality and necessity of meat eating, or naturalise the slaughter of animals—is key to understanding ethics. Others have noted the specific resonances between veganism as a practice and Foucault's notion of ethics, for example, Chloë Taylor's exploration of Foucault, ethics, eating practices, and the development of "an ethico-aesthetics of the self,"[75] or Matthew Cole's historical exploration of the trajectories of vegan practice, from ethical modes of self-conduct to the transformation of habit.[76] However, I believe Foucault's discussion of asceticism as a mode of counter-conduct offers us a different way to conceptualise these practices as a specific form a resistance; these ascetic practices are forms of ethics where the subject seeks to engage in the challenge of work upon the self ("an exercise, an exercise going from easier to the more difficult, and from the more difficult to what is even more difficult"[77]) that, given the context of overwhelming societal endorsement of animal utilisation, must lead to a somewhat solitary existence by its practitioners. Perhaps usefully, Foucault offers us a way to critically examine ascetic vegan practices as a tactic or strategy, without necessarily idealising them. As I shall discuss later, the ascetic aspects of vegan practice as a form of counter-conduct are both fascinating and, simultaneously, problematic in relation to their efficacy as a political strategy.

There is a distinct aspect of vegan practice as a mode of counter-conduct that aims, or at least has the potential to, challenge a prevailing order of truth. Vegan practitioners often confront a barrage of resistance as they come into contact with knowledge systems, as scientists, doctors, and well-meaning family and friends express concern in relation to the potential dangers vegan practitioners expose themselves to by abstaining from animal products. As I described earlier, Foucault characterises counter-conduct practices in the Middle Ages as facilitating an oppositional relationship to prevailing truths, which underpin institutions and relations of domination, particularly with respect to resistant counter-readings of Scripture, and the development of eschatological beliefs and mysticism. Here, there is a resonance between these counter-conduct movements and contemporary vegan practices: for example, many vegan practitioners work directly against scientific and medical regimes of truth, which prescribe meat and dairy-based diets as essential for vitality and functioning, relying on other forms of verification, including one's own experiences and those of others.[78] In a sense, vegan practitioners run headlong into truth, particularly where personal experiences of how it might be possible to remove oneself from an animal-based diet run against prevailing public advice and medical knowledge. This does not mean that science or medical truth has no place in discourses circulated by vegan practitioners and advocates. On the contrary, vegan practitioners and vegan organisations mobilise their own medical truths, sometimes as part of a proselytizing function. There is now a wealth of data, supported with medical knowledges, available in print publications and online, which attempt to demonstrate that it is, indeed, possible to live without using animals.[79] In many respects, these mobilisations of truth mirror the trajectories of Middle Ages counter-conduct movement practices that were directed towards reading or translating the Scripture in ways that resisted dominant readings. Rather than accept research which supports continued human utilisation of animals (some of which may be actively resourced by meat and dairy industries themselves[80]), many vegan practitioners have developed an active counter-narrative, deploying facts and figures to support the possibility of living without using animals.

None of the above is necessarily suggesting that vegan practices are inherently effective strategies in attempting to challenge the war against animals. While Foucault's counter-conduct lecture is useful for trying to understand the positive political potentiality of vegan practice, it is also, simultaneously, useful for understanding the internal limits of veganism as a strategy, or at least as the form of strategy which is often articulated by practitioners and organisations. One of these limits relates to the strategy of "desertion" itself that I have described above. In some respects, the strategy plays into a fantasy that it is, indeed, possible to remove oneself from the violence of the war on animals, and that this is useful as a strategy. Pursuing a strategy of desertion through the practice of veganism does not necessarily remove the subject from the privileges delivered by that war. This is because even if one does not personally consume animal products, it is still possible, indeed likely, to benefit from the war against animals. To use myself as an example here, I personally practice veganism; however, as an Australian citizen, I live in a nation whose significant wealth and relative high per capita living standards have been won through extraordinary and intense forms of animal exploitation and violence, which generate continuing surpluses in the production of beef, lamb, seafood, and wool.[81] In this case, my role as a contributor to the war against animals does not end simply because I choose to remove myself directly from use of animal products. My decision not to be part of the war will not end the war. It is, perhaps, for this reason that veganism, at least as a strategy practised alone as a singular goal, runs the risk of functioning as a self-focused personal asceticism, rather than as a viable means to disrupt the systemic practices that constitute the war on animals. Indeed, Foucault's discussion of asceticism points out the critical limits of this strategy, including its modality as a somewhat self-absorbed address to the self, where conduct is ordered through a growing list of challenges, potentially in a spiral that turns the self against all pleasures through a kind of hyper-vigilance, self-directed surveillance and discipline:

> If you like, the Christian pastorate added to the Jewish or Greco-Roman principle of the law a further excessive and completely exorbitant element of continuous and endless obedience of one man to another. In relation to this pastoral rule of obedience, asceticism adds another exaggerated and exorbitant element. Asceticism stifles obedience through the excess of prescriptions and challenges that the individual addresses to himself. You can see that there

is a level of respect for the law. The pastorate adds to this the principle of submission and obedience to another person. Asceticism turns this around again
by making it a challenge of the exercise of the self on the self.[82]

Here, vegan ascetic tactics risk merely imposing law obsessively upon the
self, in the name of saving animals through an individual purification,
rather than as a process or discipline of creatively working towards challenging the war against animals through modes of conduct.

I certainly do not mean to suggest that there is no tactical merit in
these sort of practices. By necessity, desertion-insubordination not only
disrupts inter-subjective and institutional violence but also the epistemological violence of the governing order; it seeks to create a new set of
truths that has been unimaginable or intolerable within the terms of the
prevailing regime. In so far as veganism has the capacity to disrupt truths
and establish new rationalities, it might constitute a productive form of
counter-conduct. However, arguably, ascetic modes of veganism risk
merely reinscribing prevailing modes of truth. One of the problems that
the epistemic violence of the war on animals creates is that life without killing animals is not considered pleasurable; indeed, it is almost thought of as
a life not worth living. Arguably, ascetic modes of veganism—where personal actions to "purify" the self from the consumption of animal products
are seen as pivotal to the political project of dismantling animal exploitation—do not necessarily challenge the "truth" that human life, vitality,
and pleasure require the killing of animals. Because these ascetic modes
of veganism rely on a model of personal sacrifice, they risk confirming
a perception that living without animal killing means having to forsake
a set of pleasures towards the goal of ending violence against animals.
Such self-focused ascetic practices potentially generate fantasies of what
Matthew Calarco identifies as "ethical purity," where a belief in "good
conscience," or a kind of moral sainthood, is idealised as a goal.[83] In a
sense, we find a resonance here between veganism organised as a potentially grim form of self-sacrifice, and Middle Ages ascetic practices which
sought to deny the pleasures of the body as a mode of political resistance:

> Being an ascetic, accepting the sufferings, refusing to eat, whipping oneself,
> and taking the iron to one's own body, one's own flesh, means turning
> one's own body into the body of Christ. This identification is found in all
> the forms of asceticism, in Antiquity of course, but also in the Middle Ages.
> Recall the famous text by Suso, in which he recounts how, in the glacial cold
> of a winter morning, he flogged himself with a whip with iron hooks that

removed lumps of flesh from his body until he reached the point of tears and cried over his own body as if it is was the body of Christ.[84]

One solution to some of these dilemmas is to emphasise vegan practice as a communal activity rather than as an ascetic activity. As Foucault observes, counter-conduct movements generated mechanisms for participants to work against obedience to prevailing orders by observing relationality with others: for example, by pledging observance of rules of ethical conduct to a fellow member of a congregation rather than to the pastor or the Church.[85] In many respects, this resonates with the potentially exciting and innovative aspects of vegan practice: namely, as comprising innovative techniques for building communities where individuals establish reciprocal modes of engagement and responsibility to each other. For example, currently many vegan practitioners actively use information technologies to locate and notify others of food venues that cater to plant-based diets, and/or develop and swap techniques for creating new forms of food which replace animal-based cuisine. In these cases, forms of community can replace solitary forms of asceticism, intensifying modalities of pleasurable engagement by facilitating networking and connection with other practitioners.[86] Here, I certainly do not intend to idealise these community formations, as they are both potentially liberating and, like all communities, simultaneously generate their own forms of power relation and hierarchy: "a different governmentality with its chiefs, its rules, and its principles of obedience."[87] Certainly, we find precisely these dangers inherent in community formation amongst vegan practitioners, for example, through internalised cultures of surveillance and disciplining within alternative communities—so-called vegan policing. Arguably, such disciplinary formations are not useful in ending the war against animals, since they deflect energies away from a broader goal of ending violence towards animals, instead focusing on provisional goals of constructing apparently perfect "cruelty free" selves. Indeed, it is not clear how effective it is to simply work towards converting individuals to make changes in their consumption. Isolated individual actions, even pursued through developed counter-conduct communities, are not enough to shift mainstay global institutions and knowledge systems. As Donaldson and Kymlicka have so eloquently pointed out: "Any theory that asks people to become moral saints is doomed to be politically ineffective, and it would be naïve to expect otherwise."[88] (Surely, our political goal should be to reduce the number of animals suffering or killed or being used for human benefit, rather than aspiring to produce new vegans, as if this were the same thing.[89]) We should also be aware of the way in which

communities of vegan practitioners, like other communities, are strati-
fied and composed in ways that rehearse, reproduce, and reiterate existing forms of class and status stratification along lines of division including
gender, race, sexuality, and ability: Amie Breeze Harper, for example, has
provided a powerful exploration of white dominance within some vegan
communities, and the role of veganism in acculturating a particular form
of ethical white subjectivity (Harper notes sharply that the "conditions of
cruelty" that attend the production of some vegan products "help certain
USA vegans practice modern ethics"[90]).

However, in thinking through counter-conduct strategies, we might
imagine individual and collective forms of counter-conduct that do not
reproduce forms of ethics modelled on personal purity or generate alternative communities which only reproduce or invent new forms of violence. Exploring this challenge, perhaps, we need to put an end to the idea
that desertion from the war against animals involves personal sacrifice or
that ending the war on animals must focus on constructing "cruelty free"
subjectivities. It is for this reason, perhaps, that Taylor suggests that "the
Animal Liberation Movement would be well-advised to follow Foucault's
suggestion that liberation movements in general should take on ethico-
aesthetic tactics, rather than relying solely on utilitarian or deontological
moral argumentation."[91] I have been particularly interested in how vegan
practice and communities might draw upon other social movements, which
have operated on similar trajectories, including notably queer communities.
Indeed, in Foucault's later writings, he identified clearly the potential for
queer communities to not merely organise to defend their rights but to
develop innovative cultures and communities which could politically challenge and offer alternatives to heteronormativity, that is, to cultivate:

> a culture that invents ways of relating, types of existence, types of values,
> types of exchanges between individuals which are really new and are neither
> the same as, nor superimposed on, existing cultural forms. If that's possible,
> then gay culture will be not only a choice of homosexuals for homosexuals—
> it would create relations that are, at certain points, transferable to hetero-
> sexuals. We have to reverse things a bit. Rather than saying what we said at
> one time, "Let's try to re-introduce homosexuality into the general norm of
> social relations," let's say the reverse—"No! Let's escape as much as possible
> from the type of relations that society proposes for us and try to create in the
> empty space where we are new relational possibilities." By proposing a new
> relational right, we will see that nonhomosexual people can enrich their lives
> by changing their own schema of relations.[92]

Importantly, it is in this context that Foucault identifies an important driving force in the capacity of queer subculture to generate new practices and ways of being with others and "inventing new possibilities of pleasure."[93] In thinking through all of this, I wonder if there is scope to reframe vegan practices as modes of pleasurable counter-conduct? Perhaps, this might be a process of understanding how personal practices by individuals and communities to desert the war on animals might seek to invent new unimaginable pleasures; war, after all, opens up the possibilities for the pleasures of secret, tactical peace. In my view, we need to explore veganism as set of imperfect, but evolving, practices, which are situationally located forms of resistance to the war on animals. These practices use community to generate counter-cultures, but must do so in ways that do not simply generate intense disciplinary and regulatory norms through ascetic practices, nor communities of surveillance and hierarchy, but instead aim at generating new forms of relation, existence, and pleasure. The epistemic dimension of the war against animals creates the perception that life is unimaginable without the pleasures of eating and using animals; a counter-conduct movement can play an important role in generating a counter-discourse, which enables the possibility of a different truth.

TRUCE—ONE DAY WITHOUT KILLING?

The development of a viable counter-conduct movement, or set of movements, might be one tactical response to the enormous political challenge before us of ending the war against animals. As I have argued above, a human consumption politics comprising vegan counter-conduct movements has potential in addressing epistemic systems, but in itself is not enough to address the deeply embedded forms of violence that are directed towards animals. Instead, there is a need for a different range of strategies that aim to effect a fundamental disarmament. What would a ceasefire look like? Here, I would like to explore a thought experiment for a political action to effect a "truce." The inspiration for the thought experiment originates from Andrea Dworkin,[94] in a 1983 address to the Midwest Regional Conference of the National Organization for Changing Men in Minnesota, to "an audience of about 500 men, and scattered women."[95] In this context, Dworkin made a remarkable plea to her audience:

> And I want one day of respite, one day off, one day in which no new bodies are piled up, one day in which no new agony is added to the old, and I am asking you to give it to me. And how could I ask you for less—it is so

little. And how could you offer me less: it is so little. Even in wars, there are days of truce. Go and organize a truce. Stop your side for one day. I want a twenty-four-hour truce during which there is no rape.

... And on that day, that day of truce, that day when not one woman is raped, we will begin the real practice of equality, because we can't begin it before that day. Before that day it means nothing because it is nothing: it is not real; it is not true. But on that day it becomes real. And then, instead of rape we will for the first time in our lives–both men and women–begin to experience freedom.

If you have a conception of freedom that includes the existence of rape, you are wrong. You cannot change what you say you want to change. For myself, I want to experience just one day of real freedom before I die. I leave you here to do that for me and for the women whom you say you love.[96]

Here, in this remarkable address from Dworkin, we find a fascinating construction of equality and freedom premised on the cessation of systemic violence. In Dworkin's view, truce is quite explicitly not a space that promises "equality in power," nor the equitable distribution of means of violence.[97] In this context, the sort of disarmament that may eventuate as a result of truce is the necessary foundation for a set of ongoing relations that might be capable of not merely reproducing existing forms of domination, but ending war, at least momentarily, on this particular front of confrontation. In other words, while truce does not claim to seek a final answer for what peace would look like, it is one way to consider a strategic space where new allegiances might be formed and which provides the foundation for a different sort of friendship, a different politics. This is a beginning for, as Dworkin phrases it, "the real practice of equality"; a practice, not as a settled state, but a necessarily messy and difficult process of negotiation and renegotiation:

I want to talk to you about equality, what equality is and what it means. It isn't just an idea. It's not some insipid word that ends up being bullshit... Equality is a practice. It is an action. It is a way of life. It is a social practice. It is an economic practice. It is a sexual practice. It can't exist in a vacuum.[98]

Dworkin's conceptualisation of equality as a "practice" and "discipline" that emerges within spaces of temporary disarmament appears to me as extraordinarily useful for thinking about the strategies we might use to address the political challenge ahead of how we construct worlds without systemic violence towards animals. Firstly, with reference to my above discussion of Foucault and counter-conduct, there is a fascinating resonance between Dworkin's conceptualisation of equality as a practice or a discipline, and theorisation by Foucault of freedom in the context of governmental power as comprising questions of self-conduct, and conduct in relation to others (what he would later describe as "the care of the self").[99] Indeed, as I have discussed, it is precisely within the context of an analysis of counter-conduct movements that Foucault discovers experimental communities that imagine forms of radical equality, operating as it were, as an inversion of the hierarchical deeply authoritarian models of pastoral power around: "we could say that in some of these communities there was a counter-society aspect, a carnival aspect, overturning social relations and hierarchy."[100] Secondly, we find here a conception of political equality that does not imagine an ideal end state where democracy and equality might be fulfilled as a principle of organisation for institutions, but instead sees equality as a somewhat transitory project that might occur in spaces where violence is temporarily evacuated. In this sense, Dworkin's truce resonates with conceptualisations of democracy as momentary or passing, as theorised by thinkers such as Sheldon Wolin[101] and Nicholas Xenos,[102] and also appears to correspond to conceptions of democracy as always in opposition to forms of institutional power, such as suggested by Jacques Rancière.[103]

Could Dworkin's imagining of a truce be the basis for thinking about a momentary disarmament in the war against animals? What sort of tactical benefit might there be in calling for one day, just one day, where we stop killing animals? If pro-animal advocates were to tactically campaign for a one-day truce, one day without killing animals, one day when the slaughterhouses would shut down, what forms would such an action take? What might it look like and achieve?

1 November is observed internationally as World Vegan Day, a day established in 1994 to commemorate the 50th anniversary of the British Vegan Society.[104] The day serves a number of functions, including promoting the possibilities of a lifestyle that does not include consuming animal products or using the products of animals, and highlighting the need for pro-animal change, such as animal rights and welfare. The idea of truce that I am advancing both resonates with this action and, simultaneously, radically departs from it. What appeals to me about the idea of a one-day truce is that rather than focusing on consumers and the consumption of meat, a call for a one-day truce would seek to intervene directly in the production process of killing animals. In this respect, a one-day truce would seek to address inter-subjective, institutional, and epistemic fields of action. The

intervention targets the institutional level of the war against animals, in so far as it aims to challenge a systemic practice of violence within slaughter industries, rather than simply how individuals ethically respond to institutional violence through their own dietary choices. While institutional violence is the focus of this action for truce, the intervention would also impact inter-subjective practices, not only in the institutional violence of mass killing, and not only in relation to the practices of workers whose labour involves killing, but also in the global impact for markets and consumers of a break in the chain[105] of meat production (more on this below). In addition, the project of achieving a truce has the capacity to directly disrupt the fundamental epistemic elements of the war against animals. By focusing on the site of the production of killing (i.e. the slaughterhouse), and not the site of consumption, this pro-animal strategy might be able to shift the particular arrangement of truths that underpins systemic violence against animals. Firstly, the action moves the focus of action away from individual consumption choices (ascetic forms of counter-conduct) towards strategies to develop counter-conduct movements, which might impact production of death on the kill floor of the slaughterhouse. This strategy would, therefore, allow us to pry the advocacy gaze away from an obsessive focus on individual ethics, and towards the institutional and epistemic challenges of system violence. Secondly, a strategy of truce would allow pro-animal advocates to underline the fact that even "humane" killing counts as violence, and that a response to this violence, indeed, the barest imaginable "ethical" response, is to offer animals a momentary reprieve. This is potentially something that both animal rights and animal welfare groups could collaborate on, since the truce would not compromise the message of either trajectory of pro-animal discourse.

What strategies would be required to effect truce? One danger in seeking to shift critical gaze from the site of consumption to the site of production is that those who are involved in the work of killing—workers in slaughterhouses—will be targeted as the source of moral concern, rather than the institutions that design, facilitate, and profit from mass killing. And, of course, workers who are directly involved in the process of killing can themselves face extraordinary forms of exploitation and violence, including in the use of low wage precarious labour and in the relatively high rates of injury that are endemic to the industry.[106] Strategies to organise a truce would need to target industries rather than workers, as a way to avoid reforms that merely substitute one hierarchy for another by redirecting institutional violence at new targets.[107] Any plan to construct a truce will need to develop a model of alliance politics, which keeps in view the larger picture, drawing together groups who are mutually impacted by domination.[108]

Might it be possible to imagine an alliance between animal advocates and labour rights groups in relation to building a truce? One of my attractions to thinking about the role of labour in enabling a truce is the long history of labour action, which has aimed at the site of production, and the potential efficacy of this sort of action. It is worth observing that while the thought experiment I have offered on truce draws from Dworkin's plea for a day without sexual violence, there is a second history that informed my conceptualisation of truce, and this is the "general strike" as an action within the repertoire of political mobilisation by the labour movement. A strike action in the context of labour involves collective withdrawal of labour from a business or industry in order for workers to make a claim for improvements in pay or conditions.[109] Throughout the history of organised labour in the nineteenth and the twentieth centuries, the strike has proved massively useful as a tool for collective bargaining, hence the strong controls applied by the State and businesses in limiting, and often severely curtailing, the use of strike action (at present, in many countries, strike action is either effectively illegal or highly regulated). An important nineteenth-century development in the strike was the "general strike," a tactical strike coordinated between different unions to close down a range, if not all, of industries, often through rolling or successive cumulative actions:[110] in his classic study of strikes in the context of the USA, Jeremy Brecher reminds us that following the turbulent industrial actions associated with the 1877 railroad general strike, there was a stark reminder delivered of the "power of workers to virtually stop society, to counter the forces of repression, and to organise cooperative action on a vast scale."[111] Indeed, general strike action has the capacity to achieve fundamental shifts in production and social relationships: the annual 1 May (International Workers Day) general strike, implemented after the 1886 Haymarket incident in Chicago, not only achieved an annual labour public holiday for many workers internationally, but is tied closely with the fundamental success of labour movements in intervening in the value chains of capital to generate the norm of the eight-hour day.[112] While it may appear idealistic or impractical to imagine achieving such conjoined strike action, the history of the use of general strikes in the late nineteenth and the twentieth centuries would suggest that they were both practically achievable, and a potentially powerful political weapon in achieving change. A recent spate of general strikes across Europe—including the 14 November 2012 anti-austerity strikes[113]—and the success of the North American Occupy movement, suggests that even though the general strike may not have the same impact within the current structural conditions,[114] there is no reason to imagine that it will not be a feature of future protest organisation.

Historically, Marxist and anarchist movements theorised and explored the general strike as a tactic useful in achieving not only the economic effect of stopping production, but also the political impact of shifting social relations of power (in the case of the anarchists, for example, faith was placed in the general strike as the mechanism by which both massive social change would occur, and direct worker control of production would be facilitated[115]). Rosa Luxemburg makes explicit this multifaceted dimension of the general strike, which simultaneously, as it were, "reflects all the phases of the political and economic struggle":

> Political and economic strikes, mass strikes and partial strikes, demonstrative strikes and fighting strikes, general strikes of individual branches of industry and general strikes in individual towns, peaceful wage struggles and street massacres, barricade fighting—all these run through one another, run side by side, cross one another, flow in and over one another—it is a ceaselessly moving, changing sea of phenomena.[116]

It is, perhaps, the capacity of the general strike to shift social relations and politicise that is fascinating, particularly in shifting the political allegiances and discourses of a social order, that demonstrates its potential to serve a pedagogic function and operate at an epistemic level. Indeed, in Spivak's 2014 essay on general strikes, the philosopher observes that this transformative dimension is precisely what is politically useful:

> Marx wanted to bring about an epistemological change in workers so they would know that they were, together, "the agent of production," and that if they stopped, then production stopped. This basic idea had always been implicit in any notion of the general strike, but that it involved a specific change of mind in workers had not been so clearly perceived.[117]

The effect of such a transformation potentially reverberates across a whole society, with a range of actors reunderstanding their relationship to power, even if the action should fail to materialise change. It is, perhaps, for this reason that Gilles Deleuze and Félix Guattari, in their short appraisal of the May 1968 insurrection in France, point to a schism or scar that is yet to be resolved, and thus generates the impetus for a change to come: "The children of May '68, you can run into them all over the place…they know perfectly well that nothing today corresponds to their subjectivity, to their potential energy."[118] In a sense, the potential of the general strike is to reshape political aspirations, allow individuals to reassess their own positionality within societies, and to reappraise social relationships.

The idea of organising a one-day truce in the war against animals, one day without killing, modelled on the protest form of the general strike, appears to me to be compelling as an aspirational goal. In part, this is because of the material capacity of such an action to genuinely effect production, perhaps, to actually enable one day without killing. The idea of stopping animal killing through labour action is ambitious, but not unachievable. Indeed, we have recent evidence of the potential of labour action to stop production. In mid-2015, rolling strikes occurred across Iceland, drawing in a large number of unions, and affecting a large number of industries.[119] One intriguing facet of the campaign was the impact of the strike by veterinarians, which extended for more than a month.[120] As food regulations require a veterinarian to be present during animal slaughter, the strike action led to a halt in animal killing in abattoirs, in turn leading to a threatened "meat shortage."[121] Media responses to the "meat crisis" were extraordinary (if not comic): one article ("Iceland Running out of Burgers as Vet Strike Causes Meat Crisis") suggested that "carnivores… [were]…clearing out supermarkets to hoard patties in their freezers."[122] The strike ended up compromising the supply of beef, chicken, and pork across Iceland; a manager of the Bónus supermarket chain is quoted in early May 2015 lamenting: "Fresh chicken is unavailable and as of next weekend the frozen chicken will be finished. Beef will also be finished and so we will only be able to offer fish, lamb and pork for the BBQ."[123] The fast-food chain KFC was threatened with a nationwide closure, with the managing director exclaiming: "This is unacceptable. We have eight restaurants and we're rather specialized since we only serve chicken. We might have to close if nothing changes."[124] Meat producers, reeling from the unplanned interruption in the chain of production, began paradoxically citing their concern for animal welfare, with one slaughterhouse CEO proclaiming: "It has certainly started to affect innocent animals… When the crowded conditions at [poultry] breeding centers have reached a certain limit it affects the wellbeing of the birds … I consider it a violation of the law on animal welfare."[125] Here, we find an extraordinary documentation of the potential for labour action to deeply compromise the production of animal death, and, in turn, to deliver significant measurable supply chain effects in relation to the production and consumption of animal-based products. These effects are, of course, short term; however, they are deep in their impact. This sort of industrial dispute need not be thought of as unlikely to be repeated; the Iceland strike could be repeated in other contexts. Indeed, in 2014, similar strike actions by veterinarians and hygiene inspectors threated to compromise meat production in the UK (where reporting suggested at the time that the "strike may well clear supermarket shelves and butchers' shops of meat and threaten summer barbecues"[126]).

Using this sort of industrial action as a model, it is clear that working towards a one-day truce in the war against animals would require some very different campaigning by animal advocates. An alliance approach would certainly be required, perhaps working to support labour organisations to organise slaughterhouse staff, including migrant and precarious workers and their communities, who are employed in the animal industrial complex. This sort of shared campaigning need not be as difficult to imagine as it might first appear. Temporary alliances between animal advocates and slaughterhouse workers have previously occurred,[127] and, of course, much information about blatant cruelty and breach of welfare standards within slaughterhouses often originates from alarmed workers themselves. Given that there are strong intersection between labour conditions and animal welfare outcomes (e.g. the deleterious effects of increased line speeds for both animals and human workers), there are potential points of collaboration. Perhaps, an alliance approach would require different messaging that provides opportunities to focus on both the violence of killing and the violence experienced by workers in the killing process. However, pro-animal advocates would need to substantially shift their regular forms of campaigning, suspending individual forms of consumption ethics in favour of looking towards the longer-term strategic problem of how we can realistically intervene in the production process of killing animals and reduce animal death. It certainly would not be possible to sustain these sort of alliances with labour organisations if animal advocates could not accept that slaughterhouse workers, whether by choice, ambition, or necessity, were bound to their labour; moreover, animal advocates would also need to accept that, perhaps, nonvegans might support goals that correlate with those of animal liberation movements, even if these same fellow travellers are unlikely to adopt plant-based diets. Given the massive challenge before pro-animal advocates, our strategies must necessarily be experimental, and, no doubt, many of our successes will be momentary and will face compromises and co-options. However, it seems clear to me that if we are to seriously challenge the inter-subjective, institutional, and epistemic violence of the war against animals, then we need to innovate in relation to our political strategy, and this means challenging prominent modes of engagement, and simultaneously creating new forms of engagement.

Conclusion

Cavalieri evokes Walter Benjamin's enigmatic essay "Critique of Violence" to think about animal resistance (this volume, Chap. 1). In turn, I would like to conclude by thinking about Benjamin's essay, and its relationship to the concepts of counter-conduct and truce I have described in the chapter. In a quite overt way, my proposal for truce resonates directly with the concept of "divine" violence that Benjamin proposes, a violence that aims not merely to reinscribe or reiterate the existing law, but to break the cycle of law-making itself and establish "a new historical epoch."[128] We should certainly be aware of the dangerous terrain that characterises Benjamin's rendering of divine violence, which he draws from Georges Sorel's[129] romanticised, somewhat fascistic, vision of the general strike.[130] (Derrida has related his own uneasiness around the concept of divine violence, in particular, the potential for an identification of an apparently bloodless and expiatory violence with "the gas chambers and cremation ovens"[131]). However, in a sense what animates Benjamin's concept of the general strike is the idea of political action that achieves a rupture, temporarily halting the circulation of authority and commodities, generating a fundamental shift in social relations as dreams of new social relations are cultivated. My proposal for a one-day truce in animal killing aims precisely at this epistemic effect. There is no grand programme here, nor any wild "mythic" imaginings that this truce will end the war against animals. However, one hopes that a one-day truce, more than simply fuelling public resentment towards a weekend of meat-free barbeques, might, instead, have the epistemic effect of highlighting not only the intense violence delivered to animals and humans in the slaughterhouse but also the lack of necessity for this violence: life without meat, after all, goes on, indeed, might be imagined as pleasurable.

But, it is also the possibility of peaceful community, between workers and advocates, between meat-eaters and non-meat-eaters, between humans and animals, that is, perhaps, most appealing. Benjamin

introduces in his essay useful concepts for understanding violence—
"law-making," "law-preserving," and "divine" violence—but, simultaneously, offers a perspective on nonviolence, that is, peaceful forms of
agreement:

> Is any nonviolent resolution of conflict possible? Without doubt. The rela-
> tionships of private persons are full of examples of this. Nonviolent agree-
> ment is possible wherever a civilized outlook allows the use of unalloyed
> means of agreement. Legal and illegal means of every kind that are all the
> same violent may be confronted with nonviolent ones as unalloyed means.
> Courtesy, sympathy, peaceableness, trust, and whatever else might here be
> mentioned, are their subjective preconditions.[132]

In a strange sense, Benjamin returns to the problem of conduct that occupied much of my discussion in this chapter. In the face of the intense hostility that characterises the war against animals, where modes of conduct,
knowledge systems, huge infrastructural and institutional deployments,
vast economies comprising inter-linked global value chains, and legal
norms and sovereign orders all conspire to enforce the "normality" of
exposing animals to ceaseless violence, it is our modes of counter-conduct
that can generate the possibilities for peace, for conduct to be agreed
through unalloyed means. We can choose to conduct ourselves with
regard to each other, and with other animals, in ways that defy the animal
industrial complex; our modes of counter-conduct can create pleasurable
forms of community, resist applied use values, and innovate with regard to
what relationships look like. Importantly, we do not need to be contained
by the rationality that monotonously circulates and underpins this massive
system of violence: "We do not want this truth. We do not want to be held
in this system of truth."

Notes

1. Michel Foucault, *Security, Territory, Population: Lectures at the Collège de France, 1977–1978* (London: Palgrave Macmillan, 2007), 191–226.
2. Foucault, *Security, Territory, Population*, 216.
3. Foucault, *Security, Territory, Population*, 216n. The full note begins with the following: "[If I have emphasised] these tactical elements that gave precise and recurrent forms to pastoral insubordinations, it is not in any way so as to suggest that it is a matter of internal struggles, endogenous contradictions, pastoral power devouring itself or encountering the limits

and barriers of its operations. It is in order to identify the 'points of entry' ('*les entrées*') through which processes, conflicts, and transformations— which concern the status of women, the development of a market economy, the decoupling of the urban and rural economies, the raising or extinction of feudal rent, the status of urban wage-earners, the spread of literacy—can enter into the field of the exercise of the pastorate, not to be transcribed, translated, and reflected there, but to carry out divisions, valorizations, disqualifications, rehabilitations, and redistributions of every kind. (…)."

In some respects, it is worth noting the similarity of Foucault's perspective on politics and tactics with respect to truth and power, and its resonance with other accounts, notably Judith Butler and her consideration of parody as a tactic of resistance within the order of normative gender construction: "Parody by itself is not subversive, and there must be a way to understand what makes certain kinds of parodic repetitions effectively disruptive, truly troubling, and which repetitions become domesticated and recirculated as instruments of cultural hegemony. A typology of actions would clearly not suffice, for parodic displacement, indeed, parodic laughter, depends on a context and reception in which subversive confusions can be fostered. What performance where will invert the inner/outer distinction and compel a radical rethinking of the psychological presuppositions of gender identity and sexuality? What performance where will compel a reconsideration of the place and stability of the masculine and the feminine? And what kind of gender performance will enact and reveal the performativity of gender itself in a way that destabilizes the naturalized categories of identity and desire." Judith Butler, *Gender Trouble: Feminism and the Subversion of Identity* (New York and London: Routledge, 2007), 189.

4. I draw this phrase from Barbara Noske and Richard Twine's utilisation. See Barbara Noske. *Beyond Boundaries: Humans and Animals* (Montreal: Black Rose Books, 1997); and Richard Twine. "Revealing the 'Animal-Industrial Complex'—A Concept & Method for Critical Animal Studies?" *Journal for Critical Animal Studies* 10.1 (2012): 12–39.

5. I am reminded here of a section of Marie Luise Knott's work on Hannah Arendt and the description of Arendt's view that intellectual endeavour must confront truth that resists theorisation of unimaginable horror in a way that resonates with William Shakespeare's famous play: "We recall that Hamlet has returned to Denmark where he is mourning the death of his father the king. He has been told that his father died of snakebite, but in reality, the king was murdered by his brother and successor, Claudius, who is also Queen Gertrude's lover. No one guesses the heinous crime and the entire court lives in the harmonious illusion of communal mourning for the dead king. The deed is so monstrous, so unimaginable, that the living

do not have language in which to conceive it. Only the ghost of Hamlet's dead father can reveal the inconceivable to him, but Hamlet cannot share what has been revealed to him with anyone. At first, he must feign madness in order to continue to live in the community, knowing the truth. A world in which the facts are unreal is a nightmare. 'Being'—living in a world that faces up to the present—is confronted by 'nonbeing'—a world that pretends to have assimilated with the false reality created by the murderers." See Marie Luise Knott, *Unlearning with Hannah Arendt* (London: Granta, 2014). 100.

6. See Dinesh Joseph Wadiwel, *The War against Animals* (Leiden/Boston: Brill/Rodopi, 2015).

7. On "symbolic" and "objective" violence, see also Slavoj Žižek. *Violence: Six Sideways Reflections* (London: Profile, 2009).

8. Guðrún Helga Sigurðardóttir. "June General Strike Looms in Iceland," *Nordic Labour Journal*, 12 May 2015, http://www.nordiclabourjournal. org/nyheter/news-2015/article.2015-05-13.4490430718.

9. See, for example, People for the Ethical Treatment of Animals, "Oliver Stone says, 'End the War on Animals,'" accessed 28 Nov. 2012, http:// www.peta.org/features/oliver-stone-says-end-the-war-on-animals.aspx. Steve Best is also recently associated with the phrase. See Steve Best, "The War Against Animals" (paper presented at the annual International Animal Rights Conference, Luxembourg, 13–16 September 2012).

10. See Liz Marshall, *The Ghosts in Our Machine*, Documentary Film (Ghosts Media Inc. 2013). See also Fusion Live and Suzette Laboy. "Animal rights activist Jo-Anne McArthur: There's a war on animals,'" *Fusion*, 19 November 2014, http://fusion.net/video/28719/animal-rights-activist-jo-anne-mcarthur-theres-a-war-on-animals/.

11. Jonathan Safran Foer, *Eating Animals* (New York: Little, Brown and Company. 2009), 33. Foer goes on to state: "If we are not given the option to live without violence, we are given the choice to centre our meals around harvest or slaughter, husbandry or war. We have chosen slaughter. We have chosen war. That's the truest version of our story of eating animals."

12. Jacques Derrida, *The Animal That Therefore I Am* (New York: Fordham University Press, 2008). 101.

13. See Thomas Hobbes, *Leviathan* (London: Everyman, 1994).

14. See Niccolò Machiavelli. *The Prince and the Discourses* (The Modern Library, 1950), especially 57–63 and 470–482.

15. Machiavelli, *The Prince and the Discourses*, 60–63.

16. See Carl Schmitt, *The Concept of the Political* (Chicago and London: University of Chicago Press, 1996).

17. See Jacques Derrida, *Politics of Friendship* (London: Verso, 2000).

18. See Achille Mbembe. "Necropolitics," *Public Culture* 15/1 (2003): 11–40.

19. See Aileen Moreton-Robinson. "Towards a New Research Agenda: Foucault, Whiteness and Indigenous Sovereignty," *Journal of Sociology* 42/4 (2006): 383–395.

20. Michel Foucault. *Society Must Be Defended: Lectures at the College de France*, 1975–1976 (London: Penguin Books, 2004). See also Michel Foucault. "Two Lectures," in *Power / Knowledge: Selected Interviews & Other Writings 1972–1977*, ed. Colin Gordon (New York: Pantheon Books, 1980) and Michel Foucault. *The Will to Knowledge. The History of Sexuality: 1* (London: Penguin Books, 1998). Aside from my discussion of these lectures in *The War against Animals*, see my earlier essay: Dinesh Joseph Wadiwel. "The War Against Animals: Domination, Law and Sovereignty," *Griffith Law Review* 18/2 (2009): 283–297.

21. Foucault, *Society Must Be Defended*, 50–51. In a separate interview, Foucault clarifies this problematic: "This is the problem I now find myself confronting. As soon as one endeavours to detach power with its techniques and procedures from the form of law within which it has been theoretically confined up until now, one is driven to ask this basic question: isn't power simply a form of warlike domination? Shouldn't one therefore conceive all problems of power in terms of relations of war? Isn't power a sort of generalised war which assumes at particular moments the forms of peace and the State? Peace would then be a form of war, and the State a means of waging it." See Michel Foucault. "Truth and Power," 109–133. 123.

22. Foucault, *Society Must Be Defended*, 95.

23. Ibid., 50.

24. Here, Foucault's starting point is Carl von Clausewitz's aphorism "war is a mere continuation of policy by other means." Foucault inverts Clausewitz in his statement "politics is war pursued by other means." See Carl Von Clausewitz, *On War*, Project Gutenberg, http://www.gutenberg.org/files/1946/1946-h/1946-h.htm.

25. For example, Moreton-Robertson, whom I mentioned above, utilises this perspective to interrogate indigenous "rights." See Moreton-Robinson, "Towards a New Research Agenda."

26. Michel Foucault, *The Will to Knowledge, The History of Sexuality: 1*, 138.

27. Ibid., 143.

28. This sort of separation is something acknowledged by different theoretical traditions, including ecofeminism. For example, see Val Plumwood's critique of discourses of human mastery: Val Plumwood, *Feminism and the Mastery of Nature* (London and New York: Routledge, 1993); see particularly "Conclusion: Changing the Master Story," 190–196. See also Val

Plumwood. *Environmental Culture: The Ecological Crisis of Reason* (London and New York: Routledge, 2002), particularly 97–122.

29. See Giorgio Agamben, *Homo Sacer: Sovereign Power and Bare Life* (Stanford: Stanford University Press. 1998).

30. Giorgio Agamben, *The Open: Man and Animal* (Stanford: Stanford University Press, 2004), 80.

31. This creates the somewhat perverse situation where, as James Stanescu describes, animals become more than just the "living dead," but an example of the "deadening of life." See James Stanescu, "Beyond Biopolitics: Animal Studies, Factory Farms, and the Advent of Deading Life," *PhaenEx* 8/2 (2013): 135–160, 151.

32. Much of the governmentality literature, like Foucault's own conceptualisation, is concerned with contemporary government as an evolution in fostering the life of human populations. See, for example, Nikolas Rose, *Powers of Freedom: Reframing Political Thought* (Port Chester, NY: Cambridge University Press, 1999) and Mitchell Dean, *Governmentality: Power and Rule in Modern Society* (London: Sage Publications, 1999), 10–16. See also Foucault's own work on governmentality, particularly Michel Foucault. *Security, Territory, Population: Lectures at the Collège de France, 1977–1978* (London: Palgrave Macmillan, 2007) and Michel Foucault, "Governmentality," in *The Foucault Effect: Studies in Governmentality, with Two Lectures and an Interview with Michel Foucault,* ed. Graham Burchell, Colin Gordon, and Peter Miller (London: Harvester Wheatsheaf, 1991), 87–104. In *The War against Animals,* I have argued that we might treat governmentality as precisely the emergence of a conjoined set of techniques, where the logic of government, derived from "pastoral power," owes its legacy to human systems of domination of animals. These techniques of power over animals happen to emerge in modernity as a refined set of techniques for governing other humans. This, perhaps, explains why, at their most extreme, the sites of extraordinary violence towards humans of the twentieth century—camps, detention centres—seem to be indistinguishable from slaughterhouses. This is something other scholars of biopolitics such as Cary Wolfe have observed. See Cary Wolfe, *Before the Law: Human and Other Animals in a Biopolitical Frame* (Chicago: Chicago University Press, 2012). See also my essay: Dinesh Joseph Wadiwel. "Cows and Sovereignty: Biopower and Animal Life." *borderlands e-Journal* 1/2 (2002), http://www.borderlands.net. au/vol1no2_2002/wadiwel_cows.html.

33. Gary Francione, *Animals, Property and the Law* (Philadelphia: Temple University Press. 2007). For a different engagement with questions of property, see John Hadley's work on property rights and environmentalism: John Hadley. "Nonhuman Animal Property: Reconciling Environmentalism and Animal Rights," *Journal of Social Philosophy* 36–3 (2005): 305–315.

34. Roberto Esposito, *Bios: Biopolitics and Philosophy* (Minneapolis: University of Minnesota Press, 2008).

35. See John Locke, *Two Treatises of Government* (Cambridge: Cambridge University Press, 2009). For a condensed version of my discussion of Locke (and the resonances between Locke and Derrida), see Dinesh Wadiwel. "The Will for Self-Preservation: Locke and Derrida on Dominion, Property and Animals," *Sub-Stance*. 43/2 (2014): 148–161.

36. Karl Marx, *Capital. Vol.1* (Harmondsworth: Penguin, 1986), 272.

37. Marx states: "the entire wisdom of those modern economists who demonstrate the eternity and harmony of existing social relations depends on this forgetting." See Karl Marx. *Preface and Introduction to A Contribution to the Critique of Political Economy* (Peking: Foreign Language Press, 1976), 11.

38. See Hobbes. *Leviathan*, and Thomas Hobbes, *De Cive: Philosophical Rudiments Concerning Government and Society* (London: J.G. for R. Royston, 1651).

39. Robert E. Goodin, Carole Pateman, and Roy Pateman. "Simian Sovereignty," *Political Theory* 25/6 (1997): 821–49.

40. Sue Donaldson and Will Kymlicka. *Zoopolis: A Political Theory of Animal Rights* (Oxford: Oxford University Press, 2011). These are not the only explorations of a concept of animal sovereignty; see also Robert Garner. "Ecology and Animal Rights: Is Sovereignty Anthropocentric?" in *Reclaiming Sovereignty.* ed. Laura Brace and John Hoffman (London: Cassell, 1997), 188–203.

41. See, particularly, Jacques Derrida, *The Beast and the Sovereign Vol. 1* (Chicago: University of Chicago Press, 2009), and Jacques Derrida, *The Beast and the Sovereign, Vol. II* (Chicago: University of Chicago Press, 2011).

42. See also Jens A. Bartelson, *A Genealogy of Sovereignty* (Cambridge and New York: Cambridge University Press, 1995).

43. On the question of the justiciability of sovereignty, see Maria Giannacopoulos. "Mabo, Tampa and the Non-Justiciability of Sovereignty," in *Our Patch: Enacting Australian Sovereignty Post-2001*, ed. Suvendrini Perera (Perth, W.A: Network Books, 2006).

44. See Jason Hribal, "'Animals Are Part of the Working Class': A Challenge to Labor History," *Labor History* 44/4 (2003): 435–453. 448–450. doi: 10.1080/0023656032000170069; and Jason Hribal, *Fear of the Animal Planet: The Hidden History of Animal Resistance* (Oakland: AK Press/ Counter Punch Books, 2010).

45. For a summary of the operaist tendency, see Sandro Mezzadra, "Italy, Operaism and Post-Operaism," in *International Encyclopedia of Revolution and Protest*, ed. Immanuel Ness (Oxford: Blackwell Publishing, 2009),

1841–1845. I have been strongly influenced in my understanding of resistance by the work of Fahim Amir, who explores operaism as a way to explain the subordination in systems of animal production. See, for example, Fahim Amir, "Zooperaismus: 'Über den Tod hinaus leisteten die Schweine Widerstand…'" (paper presented at the conference *Critique of Political Zoology*, Hamburg, 14–15 June 2013). For a discussion of the relationship of pigeons to the city, and an application of this view of resistance, see also Fahim Amir, "1000 Tauben: Vom Folgen und Fliehen, Aneignen, Stören und Besetzen." *Eurozine*, May 2013. First published in *dérive* 51 (2013).

46. I offer an expanded view of my conceptualisation of animal resistance in Dinesh Joseph Wadiwel, "Do Fish Resist?" *Cultural Studies Review*, 22(4363), 196–242.

47. Here, I make use of Clausewitz's definition of war as "an act of violence to compel our opponent to fulfil our will." See Clausewitz, *On War.*

48. Temple Grandin is famous for her designs for animal corrals with curved races: see, for example, Temple Grandin. "Race system for cattle slaughter plants with 1.5m radius curves," *Applied Animal Behaviour Science* 13 (1984/85): 295–299.

49. Ibid., 295.

50. Gayatri Chakravorty Spivak. "Can the Subaltern Speak?" in *Marxism and the Interpretation of Culture*, ed. Cary Nelson and Lawrence Grossberg (Basingstoke: Macmillan Education, 1988), 271–313.

51. Spivak, "Can the Subaltern Speak?"

52. On the international gender politics of this sort of "customary practice," see Dicle Kogacioglu, "The Tradition Effect: Framing Honor Crimes in Turkey," *differences: A Journal of Feminist Cultural Studies*, 15/2 (2004): 119–151, doi: 10.1215/10407391-15-2-118.

53. Spivak. "Can the Subaltern Speak?" 93.

54. There is a strong resonance here between Spivak's perspective and that of Edward Said in Orientalism, in understanding the dominated as constructed through an imaginary that is "arbitrarily" fabricated by the coloniser: "it is perfectly possible to argue that some distinctive objects are made by the mind, and that these objects, while appearing to exist objectively, have only a fictional reality." See Edward Said, *Orientalism* (New York: Pantheon, 1978). 54.

55. Tom Tyler. *CIFERAE: A Bestiary in Five Fingers* (Minneapolis, London: University of Minnesota Press, 2012). 21.

56. As I shall discuss below, by the 1984 lectures, Foucault had reformulated the problem of resistance to power by offering a different conceptualisation of the relationship between knowledge, government, and subjectivity.

See also Arnold I. Davidson, "In Praise of Counter-Conduct," *History of the Human Sciences* 24/4 (2011): 25–41

57. Michel Foucault, *Discipline and Punish: The Birth of the Prison* (London: Penguin Books, 1991).

58. Foucault, *The Will to Knowledge. The History of Sexuality: 1*

59. Foucault, "Governmentality." 92–93. See also Colin Gordon, "Governmental Rationality: An Introduction" in *The Foucault Effect: Studies in Governmentality with Two Lectures by and an Interview with Michel Foucault*, ed. Graham Burchell, Colin Gordon and Peter Miller (London: Harvester Wheatsheaf, London, 1991), 1–51, 11–12.

60. See Gordon, "Governmental Rationality: An Introduction." See also Mitchell Dean. *Governmentality: Power and Rule in Modern Society* (London: Sage Publications, 1999), 10–16.

61. In *The War against Animals* I provide a critique of Foucault's conceptualisation of power, which, as I argue, fails to acknowledge that it is founded upon a specific model of human domination of animals (agricultural production). I offer a condensed discussion of this critique in Dinesh Joseph Wadiwel. "Il capro di Giuda. Una rilettura della governamentalità di Foucault," *Animal Studies: Rivista italiana di antispecismo*, 11/4 (2013): 43–57.

62. As Foucault observes in the lectures, pastoral power is a "prelude to governmentality." Foucault, *Security, Territory, Population*. 184.

63. Ibid., 194–195.

64. These include "resistance, refusal, or revolt," "disobedience," "insurbordination," and "dissidence." See Foucault, *Security, Territory, Population*, 200–201. All these terms for understanding the problem are deemed inappropriate by Foucault in attempting to describe how resistance might occur within a regime of governmentality; "revolt," for example, he suggests is too closely associated with a response to sovereign power; "insubordination" too closely aligned with military practice and dissent, and while "dissidence" seemed to him potentially appropriate, the term is rejected due to the Cold War context and the over-determination of the term (Foucault remarks in the lecture that "I would rather cut my tongue out than use it"). See ibid., 200.

65. The full section reads: "what I will propose to you is the doubtless badly constructed word 'counter-conduct'—the latter having the sole advantage of allowing reference to the active sense of the word 'conduct'—counter-conduct in the sense of struggle against the processes implemented for conducting others; which is why I prefer it to 'misconduct (*inconduite*),' which only refers to the passive sense of the word, of behavior: not conducting oneself properly. And then maybe this word 'counter-conduct' enables us to avoid a certain substantification allowed by the word

'dissidence.' Because from 'dissidence' we get 'dissident,' or the other way round, it doesn't matter, in any case, dissidence is the act of one who is a dissident, and I am not sure that this substantification is very useful. I fear it may even be dangerous, for there is not much sense in saying, for example, that a mad person or a delinquent is a dissident. There is a process of sanctification or hero worship which does not seem to me of much use. On the other hand, by using the word counter-conduct, and so without having to give a sacred status to this or that person as a dissident, we can no doubt analyze the components in the way in which someone actually acts in the very general field of politics or in the very general field of power relations; it makes it possible to pick out the dimension or component of counter-conduct that may well be found in fact in delinquents, mad people, and patients. So, an analysis of this immense family of what could be called counter-conducts." See Foucault, *Security, Territory, Population.* 200–201.

66. Ibid., 205.
67. Ibid., 210.
68. Ibid., 211.
69. Ibid., 212.
70. Ibid., 213.
71. Ibid., 213.
72. Ibid., 214.
73. Ibid.,198.
74. Whether "veganism" is a practice that can be claimed as belonging to the trajectory of West is itself an important question, and beyond the scope of this chapter to fully explore. Matthew Cole, in his discussion of the history of the quarterly journal of The Vegan Society in the UK, is careful to not assert that "human plant-based lifestyles—voluntary, or involuntary—for reasons of privation, commenced in the nineteenth century, let alone 1944." See Matthew Cole. "'The Greatest Cause on Earth': The Historical Formation of Veganism as an Ethical Practice," in *The Rise of Critical Animal Studies: From Margins to Centre,* ed. Nik Taylor and Richard Twine (New York: Routledge, 2014). 203–224. However, there is a curious tendency to emphasise Western trajectories within many identifications of what constitutes veganism. This misses the important developments in practices, rationalities, and ethics that have been present in non-Western traditions, which have pursued plant-based diets, including, for example, Jainism.
75. Chloë Taylor, "Foucault and the Ethics of Eating," *Foucault Studies* 9 (2010): 71–88.
76. Cole, "'The Greatest Cause on Earth'."
77. Foucault, *Security, Territory, Population,* 205.

78. In this respect, this form of counter-conduct resonates with contemporary resistances to organised truths emanating from medical discourse: "the refusal of certain medications and certain preventive measures like vaccination, ... the refusal of a certain type of medical rationality: the attempt to constitute sorts of medical heresies around practices of medication using electricity, magnetism, herbs, and traditional medicine." Foucault, *Security, Territory, Population*, 199. In this last example, I am reminded of contemporary vegan practitioners who work against the advice of medical professionals and concerned family members to pursue a restrictive diet.

79. Consider, for example, the resources available online from the UK Vegan Society, which include resources on planning vegan diets for children and older people, and addresses concerns about aspects of nutritional intake, such as calcium and B12. See The Vegan Society, "Nutrition and Health," https://www.vegansociety.com/resources/nutrition-and-health.

80. See, for example, L.I. Lesser, C.B. Ebbeling, M. Goozner, D. Wypij, D.S. Ludwig, "Relationship between Funding Source and Conclusion among Nutrition-Related Scientific Articles," *PLoS Medicine* 4/1 (2007); and M.B. Katan, "Does Industry Sponsorship Undermine the Integrity of Nutrition Research?" *PLoS Medicine* 4/1 (2007). See also David Robinson Simon. *Meatonomics: How the Rigged Economics of Meat and Dairy Make You Consume Too Much—and How to Eat Better, Live Longer and Spend Smarter* (San Francisco: Conari Press, 2013), 9–14; and Agnes A. van der Schot and Clive Phillips. "Publication Bias in Animal Welfare Scientific Literature," *Journal of Agricultural and Environmental Ethics* 26/5 (2013), 945–958.

81. Indeed, recent international trade agreements continue to signal that Australia is continuing to position itself as a key supplier of meat and dairy in the Asia Pacific region. See Belinda Varischetti. "China live trade agreement gives unprecedented market access for Australian cattle," Australian Broadcasting Corporation, *ABC Rural*, 21 July 2015, http://www.abc.net.au/news/2015-07-20/australian-cattle-to-china/6634744.

82. Foucault, *Security, Territory, Population*, 208.

83. Matthew Calarco, *Zoographies: The Question of the Animal from Heidegger to Derrida* (New York: Columbia University Press, 2008), 136.

84. Foucault, *Security, Territory, Population*, 207.

85. As Foucault observes, this leads to remarkable experimentation in social organisation, including in modes of community, which inverted prevailing power structures: "We also find phenomena of hierarchical reversal. In these groups you have systematic reversals of hierarchy. That is to say, the most ignorant or poorest person, or someone with the lowest reputation or honor, the most debauched, the prostitute, was chosen as leader of the group... In short, we would have to study (... it's a whole problem) the

carnival practice of overturning society and the constitution of these religious groups in a form that is the exact opposite [of] the existing pastoral hierarchy. The first really will be the last, but the last will also be the first." Foucault, *Security, Territory, Population*, 211–212.

86. In some respects these communities work like secret societies that Foucault describes as a modern form of counter-conduct: "always with an aspect of the pursuit of a different form of conduct: to be led differently, by other men, and towards other objectives than those proposed by the apparent and visible official governmentality of society." Foucault, *Security, Territory, Population*, 199.

87. Ibid., 199.

88. See Donaldson and Kymlicka, *Zoopolis: A Political Theory of Animal Rights*, 253.

89. I stress here that I am not at all opposing a politics of veganism. Rather, the politics of animal liberation suggests that the main goal should be reducing the number of animals used for human benefit. Perhaps, producing more vegans might achieve this goal; however, there are a plethora of other strategies that would be more effective in meeting the goal of animal use reduction, including reducing global meat consumption by encouraging people to eat less meat, or using labour tactics to disrupt production. As I have argued in this chapter, there are a range of other reasons that developing counter-conduct cultures of veganism might be political useful, particularly in generating new cultures that innovate around knowledges for living without using animals.

90. Amie Breeze Harper, "Race as a 'Feeble Matter' in Veganism: Interrogating Whiteness, Geopolitical Privilege, and Consumption Philosophy of 'Cruelty-Free' Products," *Journal for Critical Animal Studies* 8/3 (2010): 5–27, 14. See also Richard Twine, "Ecofeminism and Veganism: Revisiting the Question of Universalism," in *Ecofeminism: Feminist Intersections with Other Animals and the Earth*, ed. Lori Gruen and Carol Adams (Bloomsbury Academic Press, 2014).

91. See Chloë Taylor's discussion of Foucault, vegetarianism, and pleasure in Taylor, "Foucault and the Ethics of Eating." Taylor states: "the Animal Liberation Movement would be well-advised to follow Foucault's suggestion that liberation movements in general should take on ethico-aesthetic tactics, rather than relying solely on utilitarian or deontological moral argumentation" (83).

92. Michel Foucault, "The Social Triumph of the Sexual Will," in *Ethics: Subjectivity and Truth*. ed. Paul Rabinow (New York: The New Press, 1997), 157–162, 160–161.

93. Michel Foucault, "Sex, Power and the Politics of Identity," in Rabinow, *Ethics: Subjectivity and Truth*, 157–162, 163–173, 165. In some respects,

Foucault's starting point for this different vision of ethics is pleasure itself: creative practices are centred on making "pleasure the crystallizing point of a new culture." Foucault, "The Social Triumph of the Sexual Will," 160.

94. Since the publication of *The War against Animals* I have been asked, perhaps with a sense of incredulity, about my interest in radical feminist voices, such as Dworkin, Susan Brownmiller, and Catharine A. Mackinnon, and whether I propose a recuperation of their approach. On one hand, I acknowledge the deep problems many of these scholars present in relation to their universalisation, particularly in relation to race, their attitudes to sex and sexuality, including in endorsing uncritical views on pornography, sex work, and sexual practices such as erotic sadomasochism, and their essentialism, including potential endorsement of transphobic positions. However, in *The War against Animals,* I read these theorists as belonging to an important legacy in feminist thought that theorised violence, and particularly a strand of theoretical engagement that was interested in understanding the incessant and nonjusticiable hostility of sexual violence as a core foundation of patriarchy as a structural system. Importantly, as my citation of Dworkin illustrates, this conceptual frame of a "war against women" creates alternative imaginings for political strategies, which must aim for "peace." As such, this trajectory of radical feminist thought remains interesting to my framework in thinking about animals.

95. Andrea Dworkin, "Take Back the Day: I Want a Twenty Four Hour Truce During Which There is No Rape," *Andrea Dworkin Online Library,* accessed January 2013, http://www.nostatusquo.com/ACLU/dworkin/ WarZoneChaptIIIE.html. In Dworkin's notes, she observes that she was presented with an opportunity to speak her truth to the audience: "this was a feminist dream-come-true. What would you say to 500 men if you could?"

96. Dworkin, "Take Back the Day."

97. Indeed, Dworkin here explicitly rejects the idea that equality might be about equivalences of violence: "It doesn't have anything at all to do with all those statements like: 'Oh, that happens to men too.' I name an abuse and I hear: 'Oh, it happens to men too.' That is not the equality we are struggling for. We could change our strategy and say: well, okay, we want equality; we'll stick something up the ass of a man every three minutes. You've never heard that from the feminist movement, because for us equality has real dignity and importance–it's not some dumb word that can be twisted and made to look stupid as if it had no real meaning." This fiery section of Dworkin's text remains relevant to thinking about the political positions that might be taken with respect to violence against animals. The somewhat insipid tendency within some strands of animal studies and posthumanism towards an agnostic or normatively uncommitted stance

towards violence against animals (e.g. "everything we eat involves violence"; or "animals do violence to each other, hence violence towards animals should not be a problem") reflects a kind of similar problem, where equality is imagined as equality with respect to the experience of violence, rather than equality as a practice and freedom generated in spaces where violence has ceased.

98. Dworkin, "Take Back the Day."

99. See Michel Foucault, *The Care of the Self: The History of Sexuality Vol. 3* (London: Penguin, 1990). See also Michel Foucault, *The Government of Self and Others: Lectures at the Collège de France 1982–1983* (New York: Palgrave Macmillan, 2010).

100. Foucault, *Security, Territory, Population*, 211–212.

101. Sheldon Wolin, "Fugitive Democracy," *Constellations* 1/ 1 (1994): 11–25, 23. For this vision of democracy as antagonistic to mainstream power relations, see Jacques Rancière. *Hatred of Democracy* (London: Verso, 2009).

102. Nicholas Xenos, "Momentary Democracy," in *Democracy and Vision: Sheldon Wolin and the Vicissitudes of the Political*, ed. A. Botwinick and W. E. Connolly (Princeton and Oxford: Princeton University Press, 2001).

103. Jacques Rancière, *Hatred of Democracy*.

104. See Claire Suddath, "A Brief History of Veganism," *Time*, 30 October 2008, http://time.com/3958070/history-of-veganism/.

105. On the never-ending chain of animal production, see Eric Schlosser. "The Chain Never Stops." *Mother Jones*, July/August 2001, http://www.motherjones.com/politics/2001/07/dangerous-meatpacking-jobs-eric-schlosser.

106. Human Rights Watch goes as far to claim that with respect to employment, there are "systematic human rights violations embedded in meat and poultry industry." Human Rights Watch, *Blood, Sweat and Fear: Worker's Rights in U.S. Meat and Poultry Plants* (New York: Human Rights Watch, 2004), 2.

107. I draw attention here to the perspective offered by Dean Spade, who has argued for the need for an alliance politics which links apparently disparate groups through a shared experience of State violence. Referring to the work of the Sylvia Rivera Law Project on the policing of undocumented migrants and the impact this has for trans immigrants, Spade observes: "Anti-immigrant sentiment was the primary motivation for these policies, though some nonimmigrant vulnerable populations have been harmed as well, and demands change from a place of shared struggle and collective analysis. Working in coalitions of groups affected by immigration enforcement, poverty, criminalization, housing insecurity, and other key sites of the maldistribution of life chances, we can aim to have no one's messaging contribute to scapegoating another vulnerable population." See Dean

Spade, *Normal Life: Administrative Violence, Critical Trans Politics, and the Limits of Law* (Brooklyn, New York: South End Press, 2011), 159.

108. Again, the work of Spade is illuminating: "At all times, attention to how work is being done, how it interacts with the broader context of neoliberal trends (surveillance, abandonment of the poor, criminalization, cooption), and whether it can impact trans survival is required… Such an analysis necessitates contextualizing law reform in a set of broader understandings of power and control and with demands for transformation rather than inclusion and recognition." Spade, *Normal Life*, 160.

109. For a US history of strike action, see Jeremy Brecher's classic study: Jeremy Brecher, *Strike!* (San Francisco: Straight Arrow Books, 1972).

110. Early examples of proposals for the general strike include William Benbow's 1832 proposal for a worker led "grand national holiday": "The preparations must begin long before the time which shall be hereafter appointed, in order that every one may be ready, and that the festival be not partial but universal." William Benbow, "Grand National Holiday, and Congress of the Productive Classes," https://www.marxists.org/history/england/chartists/benbow-congress.htm.

111. Jeremy Brecher, *Strike!*, 22.

112. See Jeremy Brecher, Strike! *Revised, Expanded and Updated Edition* (Oakland: PM Press, 2014), 33–60; Donna T. Haverty-Stacke, *America's Forgotten Holiday: May Day and Nationalism, 1867–1960* (New York: New York University Press, 2009).

113. See Carlos Ruano and Andrei Khalip, "Anti-Austerity Strikes Sweep Europe," *Reuters*, 14 November 2012, http://www.reuters.com/article/2012/11/14/us-spain-portugal-strike-idUSBRE8AD000 20121114.

114. See Jörg Nowak and Alexander Gallas. "Mass Strikes Against Austerity in Western Europe—A Strategic Assessment," Global Labour Journal 5/3 (2014).

115. In Ralph Chaplin's foundational pamphlet, "The General Strike," there is the dream of "a large scale operation in the nature of a well co-ordinated lockout of the Captains of Finance by both workers and technicians which would put an end to the profit system but leave the production and transportation of goods unimpaired." See Ralph Chaplin, The General Strike (Chicago, IL: Industrial Workers of the World, 1985), http://www.iww.org/PDF/GeneralStrike.pdf.

116. Rosa Luxemburg, The Mass Strike, chap. 4. At: https://www.marxists.org/archive/luxemburg/1906/mass-strike/ch04.htm.

117. Gayatri Chakravorty Spivak, "General Strike," Rethinking Marxism: A Journal of Economics, Culture & Society. 26/1 (2014): 9–14, 9. See also

Spivak's discussion in this essay of Antonio Gramsci and the "pedagogic" value of the general strike in politicising and transforming workers (12).

118. Gilles Deleuze and Félix Guattari, "May '68 Did Not Take Place," in Autonomia: Post Political Politics, ed. Sylvère Lotringer and Christian Marazzi (Los Angeles: Semiotext(e), 2007), 209–211, 210.

119. This included a general strike proposed for June 2015. See Sigurðardóttir, "June General Strike Looms in Iceland"; Magnús Sveinn Helgason. "Iceland Strikes Again." The Reykjavik Grapevine, 27 May 2015, http://grapevine.is/mag/articles/2015/05/27/iceland-strikes-again/; and Ned Resnikoff, "Class War Comes to Iceland." Aljazeera America, 12 June 2015, http://america.aljazeera.com/articles/2015/6/12/class-war-comes-to-iceland.html.

120. Sigurðardóttir. "June General Strike Looms in Iceland."

121. Amar Toor, "Iceland in Running Out of Meat Because of a Vet Strike," The Verge, 7 May 2015, http://www.theverge.com/2015/5/7/8564411/iceland-meat-shortage-veterinarian-strike.

122. Omar Valimarsson, "Iceland Running out of Burgers as Vet Strike Causes Meat Crisis," Bloomberg Business, 7 May 2015, http://www.bloomberg.com/news/articles/2015-05-07/iceland-running-out-of-burgers-as-vet-strike-causes-meat-crisis.

123. Eygló Svala Arnarsdótt. "Strike Leads to Shortage of Meat in Iceland." Iceland Review, 5 May 2015, http://icelandreview.com/news/2015/05/05/strike-leads-shortage-meat-iceland.

124. Jón Benediktsson. "Vet Strike Getting Serious: KFC To Close Due To Chicken Shortage." The Reykjavik Grapevine, 7 May 2015, http://grapevine.is/news/2015/05/07/vet-strike-getting-serious-kfc-to-close-due-to-chicken-shortage/.

125. Eygló Svala Arnarsdótt, "Strike Affects Animal Welfare in Iceland," Iceland Review, 24 April 2015, http://icelandreview.com/news/2015/04/24/strike-affects-animal-welfare-iceland.

126. Guardian (UK). "Where's the Beef? Strike Could Mean Meat Free Barbecues." Equities, 14 August 2014, http://www.equities.com/index.php?option=com_k2&view=newsdetail&id=56610.

    A similar meat inspector strike occurred in Ontario in 1996. See Martin Mittelstaedt, "Ontario Wants Meat Inspectors Ordered Back to Work: Government Fears Possibility of Tainted Food After Reports of Slaughterhouses Operating Illegally During Public-Service Strike," The Globe and Mail, 8 March 1996. I have quite specifically focused on labour rights activism; however, this does not preclude other forms of action that might generate a space of truce. In 2014, it is reported that some 200 Jain monks engaged in a hunger strike as part of a campaign to convert the town of Palitana into "vegetarian zone." On August 2014, the Gujarat state government acceded to the demand, purportedly creating the world's first "vegetarian city." See Andrew Buncombe. "The Vegetarian Town: They Wouldn't Hurt a Fly but the Jains Upset Palitana With Meat-free Plea," The Independent, 6 July 2014, http://www.independent.co.uk/news/world/asia/the-vegetarian-town-they-wouldnt-hurt-a-fly-but-the-jains-upset-palitana-with-meatfree-plea-9588087.html; and Shuriah Niazi, "India, The World's First Vegetarian City," Worldcrunch, 5 October 2014, http://www.worldcrunch.com/culture-society/in-india-the-world-039-s-first-vegetarian-city/india-palitana-food-meat-fish-gujarat/c3s17132.

The politics of vegetarianism in India are both fascinating and fraught: the rise of a "politicised" vegetarianism in India, and its potential use as a way to target and discriminate against non-Hindu groups mean that a much more careful analysis is required of this sort of politics, and the potential for sideways violence involved in pursuing pro-animal change. In the context of anti-Muslim violence, see Parvis Ghassem-Fachandi, Pogrom in Gujarat: Hindu Nationalism and Anti-Muslim Violence in India (Princeton: Princeton University Press, 2012), 17. See also Amit Julka and Medha. "The Politics of Modi's Vegetarianism," The Disorder of Things Blog, 11 December 2014, http://thedisorderofthings.com/2014/12/11/the-politics-of-modis-vegetarianism/. For an example of the politics of animal welfare/protection in the context of national politics in India, see Krishna N. Das, "Modi Govt Says to Push for Cow Slaughter Ban in India," Reuters, 30 March 2015, http://in.reuters.com/article/2015/03/30/india-beef-slaughter-idINKBN0MQ0LG20150330.

127. For example, the Australian Meat Industry Employees Union has publicly supported, with other organisations such as Animals Australia, the end of live animal exports from Australia. See Gonzalo Villanueva, "Mainstream crusade—how the animal rights movement boomed." The Conversation. 7 November 2012, http://theconversation.com/mainstream-crusade-how-the-animal-rights-movement-boomed-10087.

128. Walter Benjamin, "Critique of Violence," in Reflections: Essays, Aporisms, Autobiographical Writings, ed. Peter Demetz (New York: Shocken Books, 1986), 278–300, 300.

129. See Georges Sorel, Reflections on Violence (Cambridge: Cambridge University Press, 2004).

130. Spivak remarks: "We approach Sorel's work with caution since, in recommending violent general strikes as a 'myth' mobilizing the proletariat, he moved away from Marx's epistemological project and toward the heroics of fascism." See Spivak, "General Strike." See also Jack J. Roth, "The Roots of Italian Fascism: Sorel and Sorelismo," The Journal of Modern History 39/1 (1967): 30–45.

131. See Jacques Derrida, "Force of Law: The Mystical Foundation of Authority," in Deconstruction and the Possibility of Justice, ed. Drucilla Cornell, Michel Rosenfeld, and David Gray Carlson (New York: Routledge, 1992), 3–67, 62.

132. Benjamin, "Critique of Violence," 289.

Bringing the State into Animal Rights Politics

Gregory Smulewicz-Zucker

The notion that nonhuman animals are beings who deserve moral consideration and rights has found ethical justification through the innovative extensions and interpretations of deontological ethics, utilitarianism, virtue ethics, care ethics, and religious ethics. Yet, for all the robustness of the discourse on the moral status of nonhuman animals in ethical theory, political theorists have, with a few exceptions, shied away from the issue of the political status of nonhuman animals.[1] From the standpoint of current political theory, the concept of animal rights, which is widely used by theorists of animal ethics, is a misnomer. In political discourse, rights are legal rights granted, upheld, and protected by the state. In moral theory, animals are understood as rights-bearers based on an appeal to their ethical status and the obligations we have to them rather than because they have actual political rights per se. Animal rights, therefore, refers more precisely to an *ought* grounded in an ethical ideal than an *is* based on legal realities backed by the state's coercive powers. At present, animals have no enumerable rights in the sense that humans have. Animals only

G. Smulewicz-Zucker (✉)
Logos Journal, 16 West 69th St.,
New York, NY 10023, USA
e-mail: gsmulewicz@aol.com

P. Cavalieri (ed.), *Philosophy and the Politics of Animal Liberation*,
DOI 10.1057/978-1-137-52120-0_8

239

receive legal protections because of anticruelty laws or because they have the legal status of property. These limited protections are seldom enforced and do just as much to establish the allowances that the law makes for abusing animals as they do to restrict abuse. As Joan Schaffner explains, "Individual instances of gratuitous intentional cruelty against certain animals are banned, while institutionalized abuse of animals is allowed and often promoted under the law."[2] If we wish to move beyond this state of affairs, we must turn our attention to developing the kind of politics that can make animal rights a political reality. This means considering the resources at our disposal for pressuring the state to establish legal rights for animals.

In the chapter that opens this volume, Paola Cavalieri encourages us to move from the realm of morality into that of politics. Her chapter is couched in questions of theory, but it is attuned to, and draws its impetus from, the issue of practice. She compels us to ask: what kind of politics, in terms of both theory and practice, can aid us in realizing animal rights? The practice of politics means confronting differentials of power, explaining those differentials, and determining their legitimacy. Movements play a crucial role in challenging power differentials. They both work for change through activism and secure it through vigilance. Unfortunately, the animal rights movement is in a state of disarray. Though there are many individuals and organizations that are working to address the plight of animals through advocacy and activism, there is little in the way of a coherent program or strong unity of purpose. Such features are vital and foundational to the success of movements and the political change they seek to realize. There also needs to be a clear object toward which the efforts of a movement are directed. In this chapter, I argue for the development of a mass animal rights movement that is oriented toward exerting demands on the state to grant animals rights. My emphasis on a politics directed toward the state is informed by the critiques I offer of three of the most prominent alternative approaches to a politics of animal rights.

At the outset, I must stress that directing political action toward the state does not mean abandoning other useful strategies. Both theorists and activists have enjoyed considerable success without directing their attention to the state. The state, in turn, has failed animals in many respects. But, it is only the state that has the power to bestow, defend, and enforce compliance with legal rights. Further, the coercive power of the state is the only mechanism that can confront the most egregious sources of animal suffering. Animals suffer in multifarious ways at the hands of individual humans, ranging from the outright sadism of aberrant individuals to activities that are broadly deemed culturally acceptable, like hunting. As a society, we are generally appalled by the sadistic treatment of animals and condemn it. In the USA, the Animal Welfare Act is a useful resource for animal rights activists. Environmental activists can also provide good grounds for the protection of natural habitats, particularly given growing concern about global warming. While these are not ideal resources for realizing animal rights, they indicate that there is some normative basis for bringing an end to, or at least curtailing, certain forms of cruelty to animals. It is relatively easy to rally humans against the kind of outright abuse and torture of animals that imposes suffering as an end in itself. However, the larger-scale exploitation of animals for other ends, whether for food, clothing, leisure, or entertainment, is the more daunting problem confronting supporters of animal rights. In order to remedy the large-scale forms of animal exploitation, state intervention is necessary.

In liberal society, economic interests are sacrosanct. Utilizing the language of economic interest and its purported benefits is, perhaps, the single strongest argument that corporations and industries, which kill and exploit millions of animals daily, have at their disposal. The defense of economic self-interest also poses one of the major barriers to granting rights to those animals that are less likely to be exploited under capitalism or that members of the general public might deem exceptional, like dogs or cats. Corporate interests are quick to mobilize against granting rights to any animals for fear of the "slippery slope" that might follow if, for example, we grant rights to great apes. These interests are most successful in their resistance by making appeal to the liberal defense of the pursuit of economic interest. This does not mean that the language of liberalism has not been fruitfully employed on behalf of animals. Tom Regan has been a path-breaking figure in the animal rights movement in his use of Kantian moral theory to defend animals as interest-bearers with liberal *moral* rights.[3] Still, a liberal conception of *moral* rights is different from a liberal conception of *political* rights.[4] And, it is in the case of achieving political rights that I believe the power of the state becomes necessary.

The problem is that contemporary political theory is permeated by suspicion of the state. Liberals struggle over the question of state involvement in civil society. As Alasdair Cochrane explains, "liberal thinkers believe that the state should refrain from imposing one particular conception of the good on free and equal persons, but instead allow different ways of life to flourish as far as possible."[5] This position is true of the most radical liberals who support certain forms of state intervention to the most conservative liberals who want as minimalistic a state as possible. This begs the question of what might happen in practical terms when we pit a conception of the individual goods of animals against the individual goods of the person who exploits animals for the sake of their economic interests. Communitarians can support state intervention, but favor the activity and cooperation of communities.[6] They are reluctant to see the state intervene in the conception of the good of a community. It is certainly the case that certain communities would perceive the state imposition of animal rights as a violation of their conception of the good. Orthodox Marxists oppose the state as superstructural to an economic base. The state is, thus, reduced to a manifestation of the interests of capitalism. Finally, anarchists and libertarians see the state as an instrument of the oppression of some decontextualized notion of free individuality.[7]

Overcoming such antistatist tendencies in political philosophy is, in my view, one of the pressing problems for those who wish to develop a political theory of animal rights. Yet, for the purposes of this chapter, I wish to examine the kind of political practice that can be yielded by some of the most prevalent currents in political philosophy. In performing this task, I believe that I am working within the spirit of Cavalieri's opening chapter. Cavalieri is deeply concerned with the practical implications of political theory. This informs the course she takes in her critiques of liberalism and Marxism. Like Cavalieri, I begin with critiques of some of the prevailing currents of thought, but would do so from a different angle. I show how the failings of three conceptions of politics lead me to conclude that the formation of a strong animal rights movement that is oriented toward demanding state action is necessary to achieving political rights for animals.

In what follows, I begin with the posthumanist thought that has emerged out of postmodernism. I argue that this current is antipolitical and, thus, entirely useless for an animal rights politics, in both theory and practice. In the second section, I discuss some of the recent attempts to translate liberal *moral* theories of animal rights into a liberal *political* theory of animal rights. I am far more sympathetic to this enterprise because it tries to orient itself toward the state. In fact, I argue that liberal *moral* theory has played an indispensible role in the animal rights movement and believe that it will continue to do so. At the same time, I am skeptical of the kind of practical political guidance we receive if we adhere to liberalism's reliance on deliberative democracy. I believe the theory is stronger than the practice it yields. I claim that liberals who depend on deliberation inexorably find themselves falling back into moral argumentation. In the subsequent section, I turn to the politics of animal liberation groups, which focus primarily on practice over theory. I have certain sympathies with liberation activists. I think, nonetheless, that they draw unsupportable parallels between themselves and other historical movements, which harm their claims to legitimacy. I also argue that their political bearings are mainly anarchistic and focus too heavily on directing action toward economic institutions, which yield limited gains. They entirely ignore the state as a prospective positive force for animal rights. These features interfere with the capacity of these groups to realize lasting change.

In the final section, I offer a schema for a broad-based, multifaceted, unified, and organized animal rights movement. I revisit my critique of liberalism to point to those elements that can positively be derived from it. I go on to articulate the different levels at which I think such a movement ought to operate. Yet, I argue that its overall driving purpose should be to apply pressure on the state. In pursuing this argument, I make the case that directing a large-scale movement toward the state is the most practical strategy for creating a political base and developing a coherent political vision. This begs a theoretical concern about why the state ought to defend animal rights. This is a consequential question, but it is not one I can address within the scope of this chapter. I have opted, in this chapter, to gage the options on the table from the standpoint of practical offerings. As such, to the extent that it is possible, I do not engage in philosophical critiques. Instead, I focus on what the animal rights movement gains or loses if it follows the practical suggestions or implications of a theory. While there are theorists and activists whose work does not fall within the scope of my categories, I selected those currents that are most widely debated and have most clearly represented themselves as political. I am looking at general trends within a given approach. I select several representative thinkers working within those trends, but I neither exhaustively examine all the thinkers contributing to these debates nor do I exhaustively treat the argument of those I do discuss.

Postmodernism and Posthumanism: The Antipolitics of Dissolving Differences

In his 1979 lectures at the Collège de France, Michel Foucault explained that the point of his disparate subjects of inquiry (such as prisons, mental health, and sexuality) "is to show how the coupling of a set of practices and a regime of truth form an apparatus (*dispostif*) of knowledge-power that effectively marks out in reality that which does not exist and legitimately submits it to the division between true and false."[8] Following this framework of analysis, Foucault's postmodern successors have engaged in a wholesale critique of claims to knowledge. These critiques are purportedly forms of political engagement because they combat "regimes of truth" that discipline society by carving the world into categories. More recently, this foundational element of postmodernism has informed the field of posthumanist studies and its claims about the relation between humans and animals. Acknowledging the influence of Foucault on posthumanism, Cary Wolfe explains,

> The term 'posthumanism' itself seems to have worked its way into contemporary critical discourse in the humanities and social sciences during the mid-1990s, though its roots go back, in one genealogy, at least to the 1960s and pronouncements of the sort made famous by Foucault in the closing paragraph of *The Order of Things: An Archaeology of the Human Sciences*...[9]

In a recent critique of postmodernist approaches to the discussion of animal ethics, Gary Steiner has written that "while postmodernism may outwardly appear to hold the promise of dispossessing us of idealized distortions and of providing us with a more adequate grasp of reality, its real function is to leave reality and our relationship to it essentially unchanged—which is to say that it can offer us no prospect of progress in the endeavor to reduce the violence that we encounter in the world every day."[10] Though I strongly agree with Steiner's critique of postmodernism, it is somewhat surprising that, in the text I cite, Steiner makes only scarce reference to posthumanism. A critique of postmodernism is a necessary propaedeutic to a critique of posthumanism because the latter has its moorings in the former. Still, it is important to note that those who identify with posthumanism have a slightly different agenda. A viable critique of posthumanism needs to address the ways it differs from postmodernism. In this section, I explain the link between postmodernism and posthumanism and argue that the advocates of such positions make claims that are useless to a concrete politics of animal rights.

Drawing from the work of Étienne Balibar, Wolfe argues "that 'the human' is achieved by escaping or repressing not just its animal origins in nature, the biological, and the evolutionary, but more generally by transcending the bonds of materiality and embodiment altogether."[11] The posthumanist literature is rife with references to embodiment and the "human." The focus on the body supposedly recovers our repressed "animal origins." We construct ourselves as "human" through the distinctions we draw between animals and us. There is a clear affinity between this view and Jacques Derrida's discussion of our "hominization," or the evolution of our human bodies. Derrida writes,

> the dominant discourse of man on the path toward hominization imagines the animal in the most contradictory and incompatible generic terms [*espèces*]: absolute (because natural) goodness, absolute innocence, prior to good and evil, the animal without fault or defect (that would be its superiority as inferiority), but also the animal as absolute evil, cruelty, murderous savagery.[12]

The definition of what it means to be human, according to Derrida, is reliant on an imagined construct of what it means to be an animal. Yet, this "imagined" conception of the animal is bedeviled by contradictions.

Giorgio Agamben follows a path similar to Derrida when he suggests that we need to understand that the category of the "human" is defined through an opposition between the "human" and other beings, that is, animals. Agamben calls this the "anthropological machine." He writes, "Insofar as the production of man through the opposition man/animal, human/inhuman, is at stake here, the machine necessarily functions by means of an exclusion (which is also always already a capturing) and an inclusion (which is also always already an exclusion)."[13] Agamben reiterates this point when he writes, "in our culture man has always been the result of a simultaneous division and articulation of the animal and the human, in which one of the two terms of the operation was also what was at stake."[14] This opposition, according to Agamben, is essential to our self-definition. By recognizing that these distinctions are produced, we are faced with the discomfiting fact of our own indefinability: "To render inoperative the machine that governs our conception of man will therefore mean no longer to seek new—more effective or more authentic—articulations, but rather to show the central emptiness, the hiatus that—within man—separates man and animal, to risk ourselves in this emptiness: the suspension of the suspension, Shabbat of both animal and man."[15] For Agamben, the consequence of negating the opposition (and realizing that it is manufactured by the "anthropological machine") is recognition of the "emptiness" of what it means to be human.

Derrida and Agamben's approaches are generally in keeping with the postmodern emphasis on the emptiness of the definitions that we use to orient ourselves in the world. Such steps are essential to the posthumanist project of breaking apart the distinction between the "human" as a category and the "animal" as a category. Posthumanists, though, take an additional step. They are concerned with dismantling the definition of the "human" not so much to expose its failure to have a concrete referent as to break down the distinction between humans and animals. They emphasize our relatedness to animals and the ways our lives are entwined with theirs. There is a kernel of truth here. Clearly, we are tied to animals through evolution as well as by the fact that much of our societal development has relied on the use of animals. Yet, we can still appreciate that our lives are interwoven with those of animals without ignoring the fact that there are very real differences between humans and animals founded on the same biological and social realities that bind us together. It is this latter claim that posthumanists object to. The posthumanist is not merely interested in making the rather obvious point that our lives are connected to those of animals. Instead, posthumanists wish to make the additional case that the differences between humans and animals are social constructs.

Following this line of thought, Brian Massumi has recently argued that animals and humans exist along an "animal continuum." Massumi writes, "The cutoff point of the 'animal continuum' is unassignable, as is that of life. Animality and life cannot be strictly demarcated from the nonorganic. This is an inescapable consequence of affirming the logic of mutual inclusion. Calling nature's continuum of mutual inclusion 'animal' is, from this point of view, somewhat arbitrary."[16] The influence of the postmodern opposition to categorization is evident in Massumi's point. Nature is taken as an undifferentiated whole and we choose to draw lines within it. We construct the differences to serve our interests. Massumi goes so far as to suggest that even distinguishing animals from plants is a strategic choice: "the 'animal continuum' could also be called the 'plant continuum,' depending on which middle one chooses to begin from, and for which conceptually constructive strategic reasons, leading to which definitions and distinctions, to what effect. The choice is not really arbitrary. It is thoroughly pragmatic."[17] There is an implicit postmodern politics embedded in this argument evidenced by the claim that the distinctions we draw are guided by "strategic reasons," which suggests that "regimes of truth" are operative.

Massumi believes that there is a political lesson to be drawn from thinking of ourselves as part of an "animal continuum." We purportedly recover our animal selves and embodiment, which link us more closely to animals. Directing our attention to our animalism means opening ourselves up to an entirely new conception of politics. He, therefore, endorses a politics that "flourishes with noncognitive primary consciousness"[18] or "carries forward enthusiasm of the body."[19] Through this, we achieve a politics of "mutual inclusion": "The animal thinking-doing of politics refuses to recognize generic difference as foundational, precisely in order to think the singular. Its natural logic of mutual inclusion—the paradoxical logic of that which interpenetrates without losing its distinction—is designed to avoid the infernal alternative between identity and differentiation."[20] Akin to Agamben's argument that the collapse of the opposition between humans and animals leads to an "emptiness" in which we ought to "risk ourselves," Massumi argues that there are "zones of indiscernibility" along the continuum that blur the lines between humans and animals. These "zones" should be conceived of "positively, as the crucible of the emergence of the new."[21] This is why Massumi claims "Animal politics is a *politics of becoming*, even—especially—of the human."[22] Hence, politics is not tied to anything concrete—it is a flux. It entails a kind of openness to new possibilities that might emerge through the recognition of a common animal nature, and prohibits us from distinguishing ourselves from animals.

Massumi's argument echoes points made by Donna Haraway, whose essay "A Cyborg Manifesto" established her as a leading theorist of posthumanism. Massumi's claim that we exist as part of an "animal continuum" relies on a view of nature as an undifferentiated whole. Haraway similarly states,

> The world is a knot in motion. Biological and cultural determinism are both instances of misplaced concreteness—i.e., the mistake of, first, taking provisional and local category abstractions like 'nature' and 'culture' for the world and, second, mistaking potent consequences to be preexisting foundations. There are no pre-constituted subjects and objects, and no single sources, unitary actors, or final ends.[23]

Instead, Haraway proposes that we think of our relation to animals in terms of "companion species": "To knot companion and species together in encounter, in regard and respect, is to enter the world of becoming with, where *who and what are* is precisely what is at stake."[24] She goes on to state, "I am who I become with companion species, who and which make a mess out of categories in the making of kin and kind."[25] We are, therefore, part of an interrelated whole and are constituted through these relations. This "becoming" through these interrelations is another way of framing the postmodern opposition to categories because it makes "a mess out of categories." Haraway is averse to the very word "species."

For her, it entails the imposition of categories that establish differences: "*Species* reeks of race and sex; and where and when species meet, that heritage must be untied and better knots of companion species attempted within and across difference."[26] Speaking of "companion species" serves Haraway as a way to speak of the entwinement of species and to abolish the language of category distinctions.

The postmodern approach and the posthumanist approach it engenders are bound together by the task of deconstructing presupposed differences. Differences are revealed as constructions that foreclose openness to new possibilities. The problem is that, despite the fact that our differences with animals have, at times, been used to justify their abuse, real differences do exist, and glossing over them makes it impossible to speak politically about animals. There is also a marked anthropocentrism beneath these efforts to wash away distinctions. Note that it is *we* who come to recognize that *we* are animals through our encounter with animals. It is *our* "becoming" that concerns Massumi and Haraway. The postmodernists and posthumanists are, in fact, saying nothing about animals. If animal politics is a politics of becoming human, where do the animals fit in? Further, what is striking about all the talk of "anthropological machines," "animal continuums," or "companion species" is that they actually say nothing whatsoever about the implications such concepts have for discussing the political status of animals in our society.

Ignoring the differences between humans and animals ignores the differentials of power that exists between humans and animals. But posthumanists blithely dismiss biological and sociological realities as constructs. And it is precisely these realities that explain why we have power over animals in the first place. Humans have biologically evolved with capacities that enable us to subjugate animals. Our societies have developed in ways that are reliant on the control and domestication of animals. Without considering these factors, it becomes impossible to have any explanatory starting point for dealing with politics. The practice of political theory necessitates accounts of how power dynamics develop. Without these accounts, we cannot evaluate how they can be changed or whether or not those changes are legitimate. If there are no differences, there are no differences of power. Even if we entertain the notion that we have constructed differences in our favor, we are still the ones who have the power to construct the relations, and there is something about us that enables us to do this. In the absence of any actual explanation of the patent differentials of power that confront us on a daily basis, the discussion of animals in postmodern and posthuman discourse is antipolitical. It offers no account of the real forces that make the power dynamic between humans and animals possible. The animals we eat have been bred to maximize the production of animal-based commodities and to be totally dependent on the employees of factory farms. Various breeds of dogs endure breathing problems, muscle and joint pains, and are susceptible to infections because humans have bred them for traits they find aesthetically pleasing. These are not issues of constructs or a symbolic order. These are concrete expressions of the power that humans have over animals.

Distinctions matter because they are a part of the reality of the cohabitation of animals and humans. We cannot begin to develop a political theory that can address the positions animals hold in our society without first accepting that differences exist. Specific species have specific needs. Different animals suffer under different circumstances. Postmodernists and posthumanists offer no practical political solutions that address these real differences. At best, they resort to vague allusions to new possibilities. In its denial of difference, postmodernism is irrelevant to any practical political theory designed to address the condition of animals. Gary Steiner's charge that postmodernism's "real function is to leave reality and our relationship to it essentially unchanged" is absolutely correct. By rejecting solutions in favor of making the claim that differences are entirely constructed, postmodernists and posthumanists merely uphold the existing set of relations between humans and animals. They eschew any account of the institutional basis for animal oppression. Thus, their claims have no political direction.

Between Morality and Politics: The Problem of Liberalism's Commitment to Deliberation

While postmodernism cannot provide either moral or political solutions for addressing the status of animals, liberal moral theorists have provided persuasive and resilient arguments for why animals should be recognized as moral subjects. Moreover, since the demise of Marxism and reinvigoration of liberal political theory under the influence of John Rawls, liberalism has become the dominant framework for political discourse. Given the prevalence of the liberal idiom, it is understandable that, as we witness the development of political theories of animal rights, much energy has been devoted to translating liberal moral arguments for animal rights into liberal political ones.[27] I take a liberal theory of moral rights to mean that individuals have an obligation to treat one another as interest-bearers with equal moral status. A liberal political theory of rights means that the state has an obligation to grant legal rights to interest-bearing individuals and uphold those rights. In the case of humans, it has been relatively easy for liberals to make the case that moral rights serve as grounds for granting political rights. In the case of animals, the issue of agency complicates the matter and, as I will suggest, recurrently poses problems for liberal advocates of animal rights. Still, I am more worried about the overall trajectory of some liberal arguments for animal rights and the kind of politics they must endorse as a result. More specifically, I am concerned with a series of recent liberal arguments for animal rights that, I argue, culminate in recourse to deliberation as a democratic mechanism for reaching political agreements. I suggest that in making this move, animal rights politics is consigned to the practice of moral persuasion, which imposes limitations on what can be achieved.

Because my subject, throughout this chapter, is the kind of politics that can be pursued on behalf of animals, I am less concerned with laying out a theory of what it means for animals to be interest-bearers. I assume that animals have interests, and it is immaterial whether I adopt a more or less robust conception of what those interests are. I also assume that on the basis of those interests, a persuasive case can be made that animals have moral rights, that is, that they have interests that carry sufficient weight to argue that humans have obligations to them. Without the claim that animals have some morally significant interests, a liberal theory of animal rights cannot be pursued. The problem emerges with the direction one takes from there, that is, if one sets a moral argument or a political argument as one's end goal. From a practical political standpoint, having interests is insufficient to *achieving* political rights. Political rights are rights bestowed on the individual by the state, and the state assumes the responsibility to protect and enforce those rights. But political rights require an agent who can make demands on the state, whether those demands are that the state grants him or her rights, enforce his or her rights, or extend the scope of his or her rights. Agency is a crucial component of the liberal conception of rights because it enables the individual to make demands on the state.

As part of their argument for applying citizenship theory to animals, Sue Donaldson and Will Kymlicka claim that animals have agency. Animals, they argue, "exhibit various forms of agency. Animals can choose to avoid human settlement, but they can also choose to seek it out for the opportunities it offers."[28] Donaldson and Kymlicka suggest that this applies to domesticated animals, liminal animals, that is, nondomesticated animals that live among humans, and wild animals. From the outset, this claim seems to me to miss the political problem. At issue, from a practical political standpoint, is not whether animals "exhibit various forms of agency." It is whether or not animals have the specific kind of agency that enables them to orient their activities to the attainment of rights. The attainment and preservation of rights require the agency to exert demands on the state. Animals do not have this capacity. They need humans to act as agents on their behalf. In acting as political agents for animals, humans would need to make the case that the interests of animals should be translated into rights. Humans would, therefore, act to mediate between animals and the state.

Donaldson and Kymlicka's refusal to give up on liberalism's conception of agency is problematic. They insist that animals have political agency and argue that one example of the ways that animals express their agency is through political participation. Discussing movements against restrictions on the access of dogs to public space, Donaldson and Kymlicka explain, "It is humans who are doing the advocacy on behalf of themselves and their dogs. Humans are the agents here, doing the articulating and advocating. Dogs are the objects of agency, not the agents themselves. But this is to miss the way in which dogs, by their sheer presence, are advocates and agents of change."[29] All this example illustrates is a scenario in which, as Donaldson and Kymlicka note in the first sentence, humans are acting as agents on behalf of their dogs. This attachment to a notion of animal political agency betrays a theoretical commitment to agency in liberalism that does animals more political harm than good. It assumes that the plight of animals can only be politically resolved so long as we can make animal behavior appear as much like human behavior as possible. This seems to me to be an unnecessary step toward anthropomorphism that does more to restrict our political vocabulary than enhance it.

In my view, we can still adhere to liberalism and admit that there are certain kinds of agency that animals simply do not have. There is no reason to conclude that animals should not receive rights because they do not exhibit the kind of agency necessary to demand rights. Many humans are unable to exercise this kind of agency. Children, elderly dependents, and the disabled have rights, but do not act as agents in the political process. They rely on others to act as agents on their behalf. I do not see why, in principle, a similar situation should not apply to animals. For example, when Steven Wise goes to court to argue that Hercules and Leo, two chimpanzees, have rights, he is acting as Leo and Hercules' agent.[30] Recognizing that animals need humans to act as their agents in the sphere of politics does nothing to detract from arguments that animals should have rights. But what Donaldson and Kymlicka are after is something much stronger than rights. They want animals to be citizens. They argue, "entering into relations of citizenship is, at least in part, entering into relationships that involve facilitating the agency of our co-citizens, at all stages of their life course and at all levels of mental competence."[31] Though I do not see how this bypasses the fact that humans would still have to act as agents for animals, I understand why Donaldson and Kymlicka feel the need to make this move. Framing their approach in terms of relations of citizenship makes sense insofar as Donaldson and Kymlicka are trying to develop a theory of political obligations rather than one of moral obligations. If we argue that animals rely on humans to act as their agents, we are not explaining why humans have any political obligation to do so. By making animals citizens, Donaldson and Kymlicka can argue that citizens have obligations to one another.

Donaldson and Kymlicka's strategy is innovative in its move from the realm of moral obligation to that of political obligation. Yet, it is important to recognize that we often place greater stock on the power of the state to protect citizens from abuse than in the abilities of our fellow citizens. The state is compelled to intervene when a parent abuses a child. By intervening, the state asserts that its power to protect the rights of a child is greater than the power of the parent to violate their child's rights. The legitimacy of state intervention in such scenarios has widespread acceptance and feels unthreatening because we already acknowledge that children are citizens. The difference in the case of animals is that they are not citizens, so we still need to gain acceptance for the idea that they should become ones. This is a considerable hurdle. Moreover, if animals became citizens, we would have to envision far stronger forms of state intervention to enforce adherence. With the exception of most companion animals, the vast majority of animals are not the wards of caregivers who feel some degree of moral responsibility. On the contrary, most animals are the wards of the different industries that own them and, currently, have the legal right to exploit them. State intervention in the actions of an abusive parent entails a much more limited account of intervention than state intervention in industry. Yet, liberals, in general, are cautious about articulating the kind of robust role for the state that would legitimate its intervention in industry.[32]

Given the liberal commitment to avoiding strong state intervention and limiting the state's use of its coercive powers, liberals must depend on widespread agreement among citizens to institute change. Kimberly K. Smith stresses, "liberal theory should focus less on the coercive power of the state and more on its role in facilitating the emergence of creative, diverse civil society—a society that supports the welfare of *all* its members."[33] In this vein, liberal theory has increasingly relied on notions of public deliberation as a means for citizens to reach agreements on political issues and to allow for the diverse exchange of ideas in civil society. The state merely serves to guarantee a space for free deliberation and to give force to the political decisions that citizens agree on through deliberation. Consequently, the problem of animal agency, once again, arises. Because animals do not have the capacity to engage in deliberation, they must rely on agents who will deliberate on their behalf. However, human agents can only argue that animals have interests that warrant legal rights by making recourse to arguments for the moral rights of animals. Animals do not have any political rights that can be built upon. The idea that animals have moral rights is the only available starting point for making the further argument that they should have political rights. Hence, the burden is carried by the strength of the moral arguments that citizens make to one another. Because deliberative theories are invested in the achievement of widespread acceptance or agreement, arguments must be of sufficient strength to persuade the many humans who abuse or exploit animals to surrender that power. Such arguments would also have to persuade the many humans who enjoy the outcomes of animal abuse, such as being able to purchase a fur coat, to surrender their capacity to do so.

Those who advocate the use of deliberation as a mechanism for democratic political change are reluctant to endorse the use of state force to compel individuals to give up their powers. Liberal arguments for animal political rights must, therefore, place all of their hopes for success in the persuasive power of moral arguments. Donaldson and Kymlicka admit, "It is asking too much of moral arguments to expect them to overcome by themselves deeply entrenched cultural assumptions and the powerful forces of self-interest."[34] Yet, also admitting that we have little else at our disposal besides moral arguments, they add, "moral arguments should at least identify the moral resources that do exist, tapped and untapped, within our society, and should work to strengthen them. These moral resources for AR [animal rights] include ordinary folk who bond with their companion animals, dedicated members of wildlife organizations, and ecologists working for habitat conservation and restoration."[35] For the liberal committed to pursuing politics through deliberation, the case for extending legal rights to animals must always be through moral arguments. Moral arguments are powerful resources, but I am far from confident in the prospect that legal rights for animals can be realized by proffering moral arguments in democratic deliberation. While many people have been and can be persuaded by the power of moral arguments, the crucial question is whether moral arguments can confront the economic structural forces that are responsible for the widespread exploitation and domination of animals.

The reliance on moral persuasion in the context of democratic deliberation poses a serious dilemma for advocates of animal rights. It compels us to evaluate our goals and confront the horizon of the realizable. Is the aim to challenge individuals to change their views and behavior or is it to confront the structural forces that facilitate the mass exploitation of animals? Moral augmentation has a worthy track record with respect to the former, but dealing with the latter, as Donaldson and Kymlicka correctly note, seems to demand too much from the powers of moral persuasion. This problem goes to the heart of the issue of what the animal rights movement can reasonably hope to politically achieve. Do we wish to change individuals or structures? One of the unfortunate consequences of the weakening of Marxism is that liberal political theory is inadequate to the task of dealing with structural exploitation under capitalism.[36] The problem with moral argumentation is that it seems to carry insufficient force for dealing with animal exploitation under capitalism. There must, therefore, be a narrowing of the scope of what liberals hope to achieve through deliberation. This problem applies to every liberal who relies on deliberation or wishes to avoid the use of state power.

My focus on Donaldson and Kymlicka, thus far, is simply due to the fact that their liberal theory of animal rights is the most radical one in its call for animal citizenship. Yet, other liberal scholars have adopted less radical positions and opted not to endorse interspecies egalitarianism. Such scholars, like Cochrane and Robert Garner, believe that their position is more practical. Still, Cochrane believes that there is "a set of duties to animals that the state can legitimately make individuals comply with."[37] In this respect, even if Cochrane does not wish to make animals citizens, he shares Donaldson and Kymlicka's view that animals deserve legal protections. Yet, he also struggles over the question of how such rights can come into being and be enforced. Indeed, Cochrane must temper his argument for animal rights because he is fearful that a strong argument for the immediate implementation of rights holds the prospect of violating democratic procedures. He explains, "to institutionalize... animal rights immediately and in their totality has a worrying implication in terms of democratic procedures. It seems to suggest that such procedures can be and ought to be overridden."[38] Cochrane has no choice but to argue that this can only occur through democratic deliberation on the part of humans. He writes, "rather than circumventing democratic deliberation and procedures, this book speaks to them. As such, the theory hopes to *improve* citizens' and politicians' understanding of our obligations to animals. Furthermore, the arguments it offers are also hoped to *persuade* the public that the theory presented here is the best account of our political obligations to animals."[39] For Cochrane, as for Donaldson and Kymlicka, animal rights can only be achieved through moral persuasion, and moral persuasion must take place through democratic deliberation.

At this point, my claim is that liberal attempts to establish legal rights must place the actual burden of the achievement of rights on the strength of moral argumentation. Thus, what at first glance seem like political arguments for animal rights must, in practice, be moral arguments. One way of bypassing the problem of relying on persuading our fellow citizens that the moral rights of animals should be translated into legal rights is by suggesting that the state has obligations to animals. This is the strategy pursued by Robert Garner (who, like Cochrane, believes interspecies egalitarianism is too lofty a philosophical position) in his development of a theory of justice for animals: "eradicating the suffering of animals is the goal to which animals advocates ought to direct their attention. As a matter of justice, too, this goal should be an obligation of the state."[40] Garner also makes the case that the pursuit of justice justifies state intervention: "The status attached to the concept of justice means that it is a prime candidate for state enforcement."[41] I agree with Garner's turn to the state. Given the worries I have expressed about the purchase of moral argumentation, I am also somewhat sympathetic with Garner when he adds, "the fact that animals ethicists and advocates have spoken, largely, in the language of morality is, it seems to me, a mistake. If we think that animals have moral standing, that we have direct duties to them (a relatively uncontroversial claim), then it is appropriate to frame these obligations in the language of justice, because justice entails legal compulsion."[42] I am not as dismissive of "the language of morality" because, as I noted earlier, I do think moral persuasion has played an important role in winning over some humans and improving the lives of many animals. Nonetheless, I think Garner's move to talking about state enforcement is a step in the right direction.

Despite my agreements with Garner, I question the extent to which his theory can actually provide us with a practical politics that bypasses the liberal attachment to deliberation. Garner develops his theory of justice on the basis an ideal theory, which is tied to an "enhanced sentience position," and a nonideal theory, which is tied to a sentience position that aims to minimize animal suffering. Garner draws this distinction because he thinks that it is more practical to pursue the minimization of suffering even if we should truly aspire to the ideal theory. "Animals," Garner argues, "are entitled to rights, and are therefore entitled to be recipients of justice, because they have interests."[43] This equation serves as the foundation for Garner's claim that animals should be recipients of justice. Yet, in my view, it is problematic because it suggests that interests are a ground for claiming rights. As I noted earlier, agents are required to make the case that rights are warranted on the basis of interests. From this viewpoint, Garner's equation evades the practical problem of how we, as agents acting on behalf of animals in the realm of politics, should go about making the case for rights consistent with justice. Unless Garner is proposing that the state take the initiative and establish a nonideal set of rights (which I think is unlikely), there still need to be agents advocating for those rights. States fulfill obligations to citizens because citizens exert demands. The state can only have obligations to animals if humans act to exert demands on the state on behalf of animals.

Part of the justification Garner offers for his nonideal theory of justice is that it is more practical. That is, we are more likely to be amenable to the rights it entails than the rights that an ideal theory entails. As Garner explains, one of the reasons he adopts his nonideal theory is because "it may not be unrealistic to get the majority of the public to accept it."[44] But, for the sake of practical politics, there has to be some mechanism for determining which rights for animals the public would be willing to countenance. Garner says nothing about a role for the state in determining what rights to grant animals. At the same time, he says nothing of deliberation. Still, it has to be the case that *we* are deciding to accept a nonideal theory if the criterion for his nonideal theory is that it is expedient for us. In this respect, the state only serves the function of enforcing laws that we are willing to tolerate. The only mechanism I can see for determining what laws we are willing to tolerate is deliberation. Hence, in terms of practical political action for the animal rights movement, Garner's theory, like that of Donaldson and Kymlicka and Cochrane, must, ultimately, yield a politics of persuasion in a democratic public sphere. At its foundation, this argument would have to entail that animals have sufficient moral worth to make the public accept the case we are making for their legal rights. Garner suggests as much when he proposes that sentience can serve as a basis for establishing the moral worth of animals. Making the case for moral worth necessitates moral argumentation. That is, making the case that animals have moral worth must precede an argument for rights. Given that Garner wants his nonideal theory to have the public acceptance of the majority, the practical politics his argument yields must be based on the activity of making persuasive moral appeals to our fellow citizens through deliberation.

There are important substantive differences between Donaldson and Kymlicka and Cochrane and Garner. Yet, the focus of my critique is on the formal aspect of liberal arguments. Contemporary liberalism is founded on a cluster of ideas. In this case, I have emphasized its reliance on theories of agency and democratic deliberation. Liberalism finds itself in quandaries when it makes its commitment to these ideas too thin. For this reason, theorists attempting to develop a political theory of animal rights aspire to find means for reconciling animal rights with liberal principles. Yet, it is my contention that liberalism prohibits animal rights from becoming a matter of political theory because doing so would mean that it imperils itself. Granting rights to animals without subjecting the issue to public deliberation would run the risk of violating the interests of humans that liberalism is committed to protect. Liberals must, therefore, always resort to making moral cases for rights to fellow citizens in the context of a deliberative public sphere. If we wish to establish legal rights for animals, we must ask how effective these arguments can be when they must be used to confront the vested interests (whether they are strong or weak) of those who abuse animals. It is unclear how successful moral argumentation through deliberation can be when the public has to achieve some consensus that the interests of animals should trump the interests of humans. As long as we remain bound to making moral arguments to our fellow citizens, the best that liberalism can offer in terms of a practical political program for achieving animal rights is a politics of moral persuasion in the context of deliberation. Hence, what appear to be political arguments are, in reality, moral ones.

My critique of liberalism should not be taken as a condemnation of it. Even though I think the liberal arguments I have reviewed must fall back on moral persuasion, liberals take politics seriously in ways that postmodernists and posthumanists do not. They recognize that there are power differentials between humans and animals. They make arguments for rights and try to determine what those rights should be. They also consider what kinds of tools we have for realizing rights. Yet, in place of a liberal politics grounded in deliberative democracy that is directed toward persuading our fellow citizens, it is my view that we need an animal rights politics that is oriented more directly toward compelling the state to act. It bears repeating that my critique has focused on liberals whose arguments culminate in cases for a politics of democratic deliberation. I do not think that all liberals adopt this strategy. I am not offering a critique of liberalism writ large. On the contrary, I have purposefully left out liberal legal theorists, such as Steven Wise and Gary Francione. Wise and Francione are liberals of a different brand from the ones I have been critiquing. Their focus is not on persuading our fellow citizens in a deliberative public sphere, but on directing their arguments to the state. In my view, this is a kind of liberal politics that holds greater political promise.

I do not see a strategy that is more strongly oriented toward the state as an abrogation of democratic principles. There is nothing antidemocratic about directing appeals to the state as opposed to directing them toward fellow citizens. Moreover, I do not think that my critique means giving up either on deliberation or the particular arguments proffered by the theorists whose work I highlighted. Donaldson and Kymlicka, Cochrane, and Garner make useful contributions to animal rights discourse. Even if their arguments only prove to convince a single person that animals deserve rights, this is an important service. So, my critique is not a wholesale critique of their theories. Rather, it is that the kind of politics that their arguments yield is not sufficient. We need to recognize the limitations of their strategies without dispensing with them altogether. As I shall argue in the final section, the fight for animal rights must be thought of as multilayered and as including many different approaches and arguments. Thus, while I think the postmodern approach contributes nothing to an animal rights politics, I think that liberalism's offerings must be seen as limited, but, nonetheless, as an important resource. Overall, however, the direction of such a politics must be oriented toward the state.

The Misdirected Politics of the Animal Liberation Groups

For those who believe deeply in animal rights and live with the intolerable reality of the ongoing mass abuse and slaughter of animals, reliance on piecemeal efforts to improve the status of animals in our society is a source of great anguish. We wish to see change occur, but the means by which we can actively work to achieve these ends are limited. Petitions are circulated sporadically and small protests occur occasionally, but this does not have the same feel as active engagement. There is no formal, broad-based, and organized movement that maintains a public presence. There are animal advocacy groups, such as People for the Ethical Treatment of Animals (PETA) and the Humane Society, working to educate and persuade the public, but being a member of PETA is not the same as being a member of a political movement. Movements play a crucial role in translating discontent into political action. They forge solidarity so that people realize that they are not alone in their outrage at injustice. There is good reason for the supporters of animals to feel frustrated, and the means for channeling that frustration into political action are limited.

It is because of the felt absence of a coordinated political movement that groups that adopt direct action strategies, such as the Animal Liberation Front (ALF) and the Earth Liberation Front (ELF), are appealing to some animal advocates. Such organizations have focused their activities on disrupting the operation of the meat industry and the laboratories that perform testing on animal subjects. In this respect, they can boast some accomplishments, having successfully shut down some labs and small businesses that abused animals. Their spokespersons maintain that these actions serve an educative function insofar as they are meant to expose the inhumane treatment of animals to consumers. There is some truth to this as well. The exposure of the conditions under which animals live in labs and factory farms has alerted the public to what happens to animals behind the scenes, and has elicited outrage. While these actions frequently violate the law, these groups establish guidelines for their members. Animal liberation groups do not practice violence against people. Their actions are directed at the facilities and instruments that make the torture of animals possible. Thus, when they break the law, it is mainly through the destruction of property.[45]

While I am sympathetic to the effort to end the abuse of animals, I do not believe that the animal liberation groups employ an efficacious or acceptable political strategy. They have tactics, not a viable long-term strategy. Though their actions have enjoyed some, not insignificant, successes in preventing the smooth operation of industry, these can only yield short-term gains because major industries have the resources to continue to recover from small setbacks. In hoping that their actions will persuade the public, in practice, they are engaged in a politics of moral persuasion that, in principle, is not so different from the aims of the liberals they criticize and I critiqued in the previous section. Far more troubling is the romanticized self-image of the members of these groups as insurgents and their embrace of clandestine tactics. These only serve to alienate members of the public who might otherwise be sympathetic. Finally, the anarchist ethos of these groups makes them pit themselves against the state and over-simplify the relation between the state and capital. As a consequence, they neglect the only political institution that actually has the power to ensure lasting change through the rule of law.

An oft-cited justification for the activities of animal liberation groups is offered by drawing an analogy with the antislavery abolitionist movements in the USA. As Ingrid Newkirk puts it, "Today, this question of violating the laws that indemnify animal abusers looms before us just as it did in the past for those who wished to free other slaves, human slaves, from the shackles we now see on elephants in the circus, and the exploitation we now see of monkeys and rats in laboratories."[46] Newkirk's point is well taken. Charles Patterson is also not far off when he argues that animals live lives comparable to those of the inmates at the death camp at Treblinka.[47] In this vein, Maxwell Schnurer, for example, draws a parallel between the ALF and Jewish resistance fighters during the Holocaust.[48] Similarly, Gary Yourofsky writes, "Without question, ALF liberations are akin to Harriet Tubman and the Underground Railroad, which assisted in the liberation of blacks from white slave-owners."[49] I do not mean to belittle the acts of courage of the members of the Warsaw Ghetto Uprising or the Underground Railroad. However, any invocation of an historical example begs that we peer into the facts more closely. The Holocaust was aborted by the intervention of other states. Had it not been for that intervention, the mass slaughter would have continued as it did in the case of the Armenian genocide. While the Republican Party was founded as an antislavery political party, it took two years of Civil War for Lincoln to abolish slavery.

My point in briefly discussing the cases of Nazi Germany and the American South is that the actual abolition of systems that facilitated murder and enslavement required wars. In turn, the aftermath of these wars and the process of rebuilding these societies required the exercise of coercive state power. The victorious allies exerted their power to reshape the law in Germany. The US Federal government imposed the 13th, 14th, and 15th amendments on a resistant population during Reconstruction. Hence, while the analogies between animal liberation groups and the Jewish resistance or abolitionists may seem to have certain affinities in terms of the principled opposition to the law, the historical analogy is fallacious if the lesson it tries to draw is that the resistance of animal liberationists can achieve similar change. Transformations of these systems of oppression were reliant on the complete decimation of states or mass social upheaval. I am hard-pressed to see the animal liberation movement as having similar prospects. In this respect, such analogies between animal liberation groups and other historical examples of resistance are misinformed and misguided. One may, in response, cite the examples of the Civil Rights Movement and the anti-Apartheid movement. Certainly, resistance on the part of individuals and groups was an important factor. The imprisonment of leaders of these movements helped to garner international support for these movements. Yet, again, there were a host of contingent circumstances. Moreover, one of the most important and obvious differences between these movements and the animal liberation movement is that they oriented their actions toward the transformation of state policies.

The animal liberation movement is not oriented toward change at the level of the state. Its actions are directed against private economic interests. Elsewhere, I have defended the argument that the status of animals as commodities under a capitalist system helps to explain and legitimate the notion that animals are underserving of moral consideration. In targeting corporations that derive profit from the exploitation of animals, it would seem that the animal liberation movement and I are of similar minds. For example, arguing that industry should be targeted, Nicolas Atwood claims that animal liberation groups should plan "bigger, more focused actions intended to weaken or remove a vital link in an animal abuse industry. Activists must get to know their local animal industry and the role it plays at the national and local levels. An industry is made up of many different levels, from the farmers, the animal transporters, and the slaughterhouses to the processors and down to the retail end."[50] Yet, despite these acts of disruption, massive corporations are resilient. I do believe in challenging industry and acknowledge that disruption has served many social movements well.[51] However, in targeting industry there is an underlying assumption on the part of animal liberation groups that enough disruption will lead industries to reform themselves. Some industries have done so. Some clothing companies have stopped selling products made of animals. Some cosmetic companies have abandoned animal testing. These are not trivial achievements and it is unlikely that these companies would have reformed themselves without the work of liberationists or the exposure of their inhumane practices. But these companies made reforms because they were able to do so without placing their futures at risk. The industries that cause the greatest amount of suffering would imperil their existence by introducing change. In such instances, activists are confronted with the task of continually cutting off the heads of the proverbial hydra.

Despite my criticism that animal liberation groups yield short-term gains, I do think that they have, at the very least, two justifications. One is that, even they if they do not end the institutionalized abuse of animals, they can rescue individual animals and relieve them of their suffering. From a moral standpoint, this is no mean achievement. Moreover, it is not necessarily the case that a rescued animal will be replaced by another animal who will suffer the same abuse. Animal liberation organizations have shown success in forcing particular small-scale businesses that rely on animal abuse to close down.[52] Even if these kinds of actions do not lead to the closure of a large-scale factory farm, they are important. Another justification is that most animal liberation activists do this work out of conscience. Conscience should not be deprecated. It is a strong motivator for making people feel obligated to face injustice. But the larger question we must ask is whether or not this constitutes a politics. Indeed, justifying one's actions on the basis of conscience has the dangerous potential to lead to self-righteousness and the refusal to offer any objective justification for one's actions. It leads to a refusal to be held publicly accountable. Rod Preece is correct when he warns, "As a consequence of their efforts to convince others of what they find so indubitably valid, many animal rights advocates… assert their case rather than argue or explain it."[53]

I do not deny that animal liberation activists might have the right moral motives. I question whether animal liberation, as activists practice it, constitutes a political program that employs a strategy that can achieve greater gains. Politics demands a vision for different kinds of social arrangements, and it means building the kinds of institutions that can legally sustain and enforce such changes. Animal liberation activists have derived the wrong historical lessons when they cite the history of resistance. Lasting change has come through state involvement. The successes of the labor movement, for example, can be measured not only by activism, but also by the achievement of the legal right of workers to unionize and strike. This victory constrained the capacity of capital to exploit workers. Animal rights proponents need to think in these larger terms. Its theory and practice have been, for too long, couched in ethics. Not enough thought has been given to how to realize changes by means other than winning over public support through moral outrage. It seems to me that part of the problem with animal liberation activism has been that it has eschewed such a vision.

There is a troubling anarchist undercurrent that has blocked this vision. It too simply conflates the state with economic interests and is, at times, openly hostile to it. Hence, David Naguib Pellow lists the numerous examples of state repression of social movements and activists.[54] These examples miss the point entirely. The state has undeniably targeted social movements, but the social movements that have succeeded have been those that continued to pressure the state to change. Unfortunately, the refusal to see the state as anything more than an expression of the interests of the powerful has bred unduly militant and flagrantly romantic rhetoric. For example, in Steven Best's view, the animal rights movement is engaged in warfare against the state:

> Because the state is so strong in its monopolization of the means of violence, this is not a war of opposing tanks and troops, but rather a guerrilla war in which liberation soldiers disperse into anonymous cells, descend into the underground, maneuver in darkness, deploy hit-and-run sabotage strikes against property, and attempt to intimidate and vanquish their enemies.[55]

Such language does not serve animals. It assumes too simplistic an understanding of politics, and the rhetoric only alienates potential supporters. The state has been a powerful instrument serving economic interests, but it has also been made to serve interests consistent with conceptions of justice and equality. Myopia about the usefulness of the state as an entity that can stand against injustice has done more harm than good to social movements. We witnessed the deleterious effects of the romanticization of disruption in the absence of a clear set of aims in the inefficacy of Occupy Wall Street. Direct action is not an end in itself. It is a tool for a larger goal. Animal activists need a viable alternative strategy and vision for how to bring the mass abuse of animals to an end. Only once this is offered will people join a movement where they feel their activism will have some lasting effect.

BUILDING A MASS MOVEMENT AND TAKING ON THE STATE

At this point, it might prove useful to summarize the three criticisms I have developed and conclusions I have drawn: (1) postmodern and posthumanist approaches are useless for informing an animal rights politics because they fail to offer an account of power differentials, (2) the liberal approach yields a politics of moral persuasion. This is a useful strategy for winning over the support of some individuals, but is inadequate to the task of guiding a political movement that can challenge powerful interests, finally, (3) animal liberation groups do not shy away from confronting powerful economic interests, but they are ineffective in terms of a long-term political strategy. It should be clear that my criticisms have something in common. In each case, I have offered reasons why I think the absence of a politics directed toward the state limits what an animal rights politics can hope to achieve. I should make plain what I think an alternative animal rights politics, a politics oriented toward the state, would look like. In my view, its features are far milder than my talk of the need for the coercive power of the state might have made it seem. Simply put, I am suggesting that there needs to be a more robust conception of the animal rights movement such that it is able to exert significant influence to demand that the state establish animal rights. Like any other successful movement, the animal rights movement needs to unite different groups, employ different approaches, and have a coherent vision of its goals and alternatives. Moreover, like any other successful movement, it should orient itself toward compelling the state to grant rights to animals, enforce those rights, and leave open the possibility that those rights can be extended.

There are immediate practical concerns that become so obvious that it might seem I am already subject to the same criticisms I launched in the three preceding sections. Among supporters of animal rights, there is so much disagreement about questions of strategy and desirable goals that it would seem my hopes for a broad-based movement would be torn apart by competing points of view, leaving advocates of animal rights in a worse state of organizational disarray than we currently are. However, movements are compelled to face such birth pangs and there is no guarantee that they survive them. One of the fundamental functions of movements is to organize people with differing views around a coherent public position. What is required are enough people who are willing to look at the larger issues at stake and ask whether animal rights can be achieved outside of a mass movement. Members of a movement also have to determine what goals they wish to pursue, both in the short term and the long term. There has to be some agreement. Rather than necessitating the agreement of the majority of society, this agreement needs to be reached only by those who have decided to join a movement for animal rights. Nonetheless, given the extremes of opinion, there will always be people who disagree and will demand more or less from the movement. Movements do not make everyone who joins them happy. Consequently, movements are often compelled to exclude people. Movements do not require blind allegiance or silencing of criticism. They do require some fundamental consensus, a willingness among their members to postpone some aims in the name of others, and that members tolerate disagreements.

As I have tried to show by way of critique, I do think that an animal rights movement has to have an overall orientation toward exerting demands on the state and applying pressure on the state to act. In this volume, Donaldson and Kymlicka similarly support the formation of an animal rights social movement. But what differentiates my position from theirs is that their primary focus is on changing social norms. I agree that this is an important part of any movement, but it seems too narrowly focused on effecting change within civil society. In practice, this means depending heavily on the kind of moral persuasion through public deliberation that I critiqued earlier in this chapter. A vibrant social movement needs to engage in public deliberation, but it should also direct its activism toward compelling state action. This is both because the state is the only institution in society that can actualize rights and because it is the only institution that has the power to enforce the rule of law on all members of society equally. Moreover, the state is the only institution that can exercise its coercive power against vested interests. This means accepting that the state will have to intervene in the industries that profit from the abuse of animals by enforcing animal rights.

There is nothing antidemocratic about state interference with industry. In the USA, the National Labor Relations Act introduced laws that constrained industry. The Act was achieved because the labor movement had significant numbers and strength to challenge the interests of capital. This means that the animal rights movement needs to attract a growing number of supporters. It is for this reason that I am dismissive neither of moral persuasion nor of any particular theory that can persuade people that animals should have rights. So, despite my criticisms of deliberative democracy, I do believe that moral persuasion is important. The continued presentation of novel persuasive arguments needs to be ongoing. Some people will be persuaded because of their liberal principles, while others will be persuaded because of their religious ethics. In the USA, very different people supported marriage equality for very different reasons. In the end, it was their support that mattered. Still, part of the way to attract people is by having an organization that they can be a part of and that allows them to publicly express their support through participation in a movement. There is good cause for optimism that such support already exists. A recent survey shows that one-third of Americans believe that animals should have equal rights.[56] An animal rights social movement, thus, has to tap into this support and mobilize people to act.

In addition to rallying the support of the general public, movements require organizations that rally intellectuals. The animal rights movement already has important resources in this respect. The Oxford Centre for Animal Ethics, under the directorship of Rev. Andrew Linzey, has performed some of the hard work of establishing a home for intellectuals.[57] Similar, but smaller, centers and study groups have been founded at many other universities and colleges throughout the world. It is important for supporters of animal rights to establish centers and study groups in their home universities and colleges. This would enable them to found national and international conferences and journals that provide important forums for fostering communication, debate, and collaboration. Centers at institutions of higher education would be able to solicit funds from their university for conferences and journals. In addition, it is important to build international foundations that can provide research grants to established scholars and aspiring ones. Such intellectual work must also occur outside of the framework of the academy. Public policy institutes and think tanks can publish reports and develop talking points for researchers that enable them to counter the claims of opponents of animal rights. This means cultivating public intellectuals who can engage in public debate and make media appearances. There should be the ongoing publication of press releases that are regularly released to the media and stables of experts who are prepared to intervene in debate and articulate alternatives. For example, every time there are revelations about the outbreak of disease at a farming facility, policy institutes should issue statements and work to have their staff and researchers appear in the media. Regular public appearances by members of think tanks would help to garner greater public interest and persuade members of the public that animal rights is both morally just and practically achievable.

A broad-based animal rights movement also needs to be continually engaged in bringing forward legal cases. In this respect, the work of lawyers, like Steven Wise, and legal theorists, like Gary Francione, is vital. Though there are significant disagreements among legal scholars about appropriate legal strategies, such disagreements should not prevent lawyers from consistently bringing civil suits against corporations that abuse animals. Again, there are already organizations doing this work, such as, in the USA, the Animal Legal Defense Fund, which already has chapters at law schools throughout the country, or the Nonhuman Rights Project, which is run by Wise.[58] Recently, as I have already mentioned, Wise defended the case of Hercules and Leo, two chimpanzees used in locomotion studies at SUNY Stony Brook, before the Manhattan Supreme Court.[59] Such cases are indispensible for setting precedents. There is no reason why other cases should not be brought forward that, for example, engage issues like those that concern Francione in his advocacy of abolishing the property-status of animals. An intensified animal rights movement needs to direct greater attention to the institutionalized abuse of animals. Legal scholars can disagree with one another about the feasibility of the suits that particular lawyers file while still showing solidarity with one another. Bringing legal cases on behalf of the interests of animals is a way to apply pressure on the state. A valuable model to aspire to is provided by the American Civil Liberties Union (ACLU). An animal rights legal organization of comparable size can help to coordinate the action of lawyers and lend them support even if the national organization does not itself file a suit.

Citizens can pressure the state themselves through mass protests. One of the problems with the animal liberation groups I discussed in the previous section is that they work clandestinely as members of cells. This renders activism a matter of individual commitment. The self-conception of animal liberation activists as rebels does little to forge larger and transparent alliances among activists. Umbrella groups that can organize mass protests at both the local and national levels and teach members how they can participate are necessary. While PETA does provide information to its members about protests, the protests on behalf of animals have occurred mainly on a very small scale. At the same time, members of PETA offer guidance and training to supporters who wish to organize demonstrations.[60] Training and guidance for activists is helpful, but organizations that initiate mass protests remove some of the burden from individuals who might find the task of organizing a protest too intimidating. Moreover, many of the smaller animal rights organizations that exist should collaborate with one another in planning mass protests. This enables organizations to pool resources. Larger organizations should work to invigorate smaller ones and lend support to activists working at the local level.

In many respects, PETA is an important model, and some might respond that much of the work I am suggesting needs to be done has already been undertaken by PETA. This is not true. PETA's many admirable campaigns have focused primarily on advertising and promoting boycotts. Since its inception, PETA has gained growing public support through its campaigns. It has also been tremendously successful in dispensing information to the public, whether in the form of publicizing the cruel conditions under which animals live on farms or in laboratories, or circulating pamphlets that guide people in becoming more ethical consumers of products. Such work is crucial to a movement. As a consequence of the publicity it has gained, PETA has come to be regarded as a legitimate representative of the interests of animals and their supporters. It is, nonetheless, a mistake to see PETA as occupying any kind of leadership position in a powerful animal rights movement. It offers little in the way of a larger alternative vision for society, which is something that a mass movement needs. More importantly, in its focus on exposing the practices of industry and persuading consumers, its practical politics treats moral persuasion as its primary goal. A successful movement should incorporate advocacy groups, yet it requires the kind of mass organization that orients their work toward compelling state action.

In general, supporters of animal rights have shied away from creating a powerful presence in the electoral process. While organizations sometimes encourage voters to contact their representatives, more work needs to be done in influencing the electoral process. Through their membership, mass organizations have the power to determine what kind of candidates can run for office and which candidates win office. Mass organizations can position themselves as a force in the electoral process by endorsing candidates who take positions favoring animal rights. An organization that has the power to mobilize its members to give or withhold their support of candidate has a real political presence. A politics of animal rights requires making animal rights a voting issue. This means paying greater attention to political candidates at the local and national levels and reaching out to their potential constituents. The same should be done with respect to politicians who are already in office. Tracking the records of politicians on animal rights issues can be used to make politicians nervous if they are up for reelection. This means forcing politicians who have been indifferent to animal rights to take a stance and threatening to unseat those with bad records. In order to make animal rights a concern for politicians, there has to be an animal rights lobby that can contribute funds to campaigns.

This is a way for the animal rights movement to engage politicians and legislators in a far more robust fashion than by signing petitions.

New York City's 2013 Democratic mayoral primary election offers an instructive case study in how the condition of animals can become a voting issue. Nonprofit organizations working to ban horse-drawn carriages in New York City helped to ensure that New York City Council Speaker Christine Quinn, an established politician who was a clear frontrunner in the race, was not elected as the Democratic candidate. In her place, an unlikely candidate, New York City's Public Advocate Bill de Blasio, a supporter of the ban, won the election. In addition to promising to end the carriage industry, de Blasio promised to pass legislation to ban the sale of puppies from puppy mills in pet stores. Mayor de Blasio certainly did not win simply because of his opposition to the carriage industry or puppy mills. Nevertheless, because de Blasio took a position on these issues, they became matters that could influence the decisions of voters. De Blasio's campaign showed that the plight of animals could be made an electoral issue. De Blasio delivered on his promise to ban the sale of puppies from puppy mills in pet stores with the passage of a bill in the City Council. While legislation to ban the carriage industry has not (at the time of the writing of this chapter) been introduced to the New York City Council, de Blasio has stated that he still intends to do so, and the same organizations that assisted him in winning his election are continuing to pressure him to act. This is a concrete case in which what was once a moral argument has been elevated to a political argument. It did not rely on agreement through democratic deliberation among members of the public. Rather, it has become an actual democratic conflict in which interests are pitted against one another.

An animal rights movement that has a real political presence is one that has the power to challenge the interests of those who benefit from the abuse of animals in the sphere of actual democratic politics. My argument for a movement that demands state intervention is entirely consistent with democratic practice. It is, however, guided by a conception of democracy as something that happens through political engagement and occurs when interests compete. The prospects for small- and large-scale victories are completely dependent on building a mass movement that is composed of many different elements and tactics. I have suggested some of the necessary features of a mass movement for animal rights. Many of the resources already exist. There are organizations that can be enhanced and replicated that already exist within the animal rights movement. There are models to turn to for strategies. It is a mistake to think that democratic action means achieving agreement or that political action means direct action. The practice of politics is far more robust and complex. It requires working at many different levels and employing many different tactics. Limiting activity to a particular sphere or particular tactic only serves to consign and constrain the animal rights movement. Realizing animal rights means having a mass movement that has an overall orientation toward the institution that is responsible for rights, the state.

NOTES

1. Hereafter, I will be referring to nonhuman animals as animals.
2. Joan E. Schaffner, *An Introduction to Animals and the Law* (London: Palgrave Macmillan, 2012), 192.
3. The classic statement of this position is Tom Regan, *The Case for Animal Rights* (Berkeley: University of California Press, 2004).
4. I find Alasdair Cochrane's explanation of the distinction very useful and draw from it. See Cochrane's *Animal Rights Without Liberation: Applied Ethics and Human Obligations* (New York: Columbia University Press, 2012), 13–14.
5. Alasdair Cochrane, *An Introduction to Animals and Political Theory* (London: Palgrave Macmillan, 2010), 7.
6. I draw from Alasdair Cochrane's useful discussion of communitarianism in ibid., 72–92.
7. For my critique of contemporary anarchist thought, see Gregory Smulewicz-Zucker, "Illusory Alternatives: Neoanarchism's Disengaged and Reactionary Leftism" in *Radical Intellectuals and Subversion of Progressive Politics*, ed. Gregory Smulewicz-Zucker and Michael J. Thompson (London: Palgrave Macmillan, 2015).
8. Michel Foucault, *The Birth of Biopolitics: Lectures at the Collège de France, 1978–1979*, ed. Michael Senellart, trans. Graham Burchell (New York: Picador, 2008), 19.
9. Cary Wolfe, *What is Posthumanism?* (Minneapolis: University of Minnesota Press, 2010), xii.
10. Gary Steiner, *Animals and the Limits of Postmodernism* (New York: Columbia University Press, 2013), 3.
11. Wolfe, *What is Posthumanism?*, xv.
12. Jacques Derrida, *The Animal That Therefore I Am*, ed. Marie-Louise Mallet, trans. David Wills (New York: Fordham University Press, 2008), 64.
13. Giorgio Agamben, *The Open: Man and Animal*, trans. Kevin Attell (Stanford: Stanford University Press, 2004), 37.
14. Ibid., 92.
15. Ibid.

16. Brian Massumi, *What Animals Teach Us about Politics* (Durham: Duke University Press, 2014), 52.
17. Ibid., 53.
18. Ibid., 40.
19. Ibid., 41.
20. Ibid., 49.
21. Ibid., 50.
22. Ibid.
23. Donna Haraway, *The Companion Species Manifesto: Dogs, People, and Significant Otherness* (Chicago: Prickly Paradigm Press, 2003), 6.
24. Ibid., 19.
25. Ibid.
26. Ibid., 18.
27. Again, I am building off the distinction in Cochrane, *Animal Rights Without Liberation*, 13–14.
28. Sue Donaldson and Will Kymlicka, *Zoopolis: A Political Theory of Animal Rights* (Oxford: Oxford University Press, 2011), 65.
29. Ibid., 113–114.
30. For more information on this case, see "Do chimps have rights? NYC judge weighs fate of Leo and Hercules," CBS News, May 27, 2015, accessed July 15, 2015, http://www.cbsnews.com/news/chimpanzee-rights-new-york-court-to-weigh-fate-of-two-animals-kept-at-stony-brook-university/
31. Donaldson and Kymlicka, *Zoopolis*, 60.
32. Kimberly K. Smith expresses her concern with the expansion of state power in *Governing Animals: Animal Welfare and the Liberal State* (Oxford: Oxford University Press, 2012), 165.
33. Ibid.,166.
34. Donaldson and Kymlicka, *Zoopolis*, 256.
35. Ibid., 256–257.
36. I have argued elsewhere that capitalist exploitation is an animal rights problem. See my "The Problem with Commodifying Animals" in *Strangers to Nature: Animal Lives and Human Ethics*, ed. Gregory Smulewicz-Zucker (Lanham: Lexington books, 2012).
37. Cochrane, *Animal Rights Without Liberation*, 207.
38. Ibid., 15.
39. Ibid., 207.
40. Robert Garner, *A Theory of Justice for Animals: Animal Rights in a Nonideal World* (Oxford: Oxford University Press, 2013), 168.
41. Ibid., 59.
42. Ibid.
43. Ibid., 106.
44. Ibid., 139.

45. Note, for example, "Animal Liberation Front Guidelines" in *Terrorists or Freedom Fighters? Reflections on the Liberation of Animals*, ed. Steven Best and Anthony J. Nocella II (New York: Lantern Books, 2004), 8.

46. Ingrid Newkirk, "Afterword: The ALF: Who, Why, and What?" in ibid., 342.

47. For Patterson's argument, see his *Eternal Treblinka: Our Treatment of Animals and the Holocaust* (New York: Lantern Books, 2002).

48. Maxell Schnurer, "At the Gates of Hell: The ALF and the Legacy of Holocaust Resistance" in Best and Nocella, *Terrorists or Freedom Fighters?*, 106.

49. Gary Yourofsky, "Abolition, Liberation, Freedom: Coming to a Fur Farm Near You" in ibid., 130.

50. Nicolas Atwood, "Revolutionary Process and the ALF" in ibid., 273–274.

51. For an argument about the efficacy of disruption as a strategy for social movements, see Frances Fox Piven, *Challenging Authority: How Ordinary People Change America* (Lanham: Rowman & Littlefield Publishers, Inc., 2006).

52. Note, for example, the cases of the successful closure of Consort Kennels and Hillgrove Cat Farm, discussed by Kevin Jonas in "Bricks and Bullhorns" in Best and Nocella, *Terrorists or Freedom Fighters?*, 264–266.

53. Rod Preece, "Ideology in Animals Rights Advocacy: Sound Ethics, Dubious Practices" in Smulewicz-Zucker, *Strangers to Nature*, 131.

54. Note the discussion in David Naguib Pellow, *Total Liberation: The Power and Promise of Animal Rights and the Radical Earth Movement* (Minneapolis: University of Minnesota Press, 2014), 163–209.

55. Steven Best, "It's War! The Escalating Battle Between Activists and the Corporate-State Complex" in Best and Nocella, *Terrorists or Freedom Fighters?* 303.

56. Tanya Lewis, "Should animals have the same rights as people?" CBS News, May 26, 2015, accessed July 15, 2015, http://www.cbsnews.com/news/should-animals-have-the-same-rights-as-people/

57. More information can be found at the Centre's Web site: Oxford Centre for Animal Ethics, accessed July 15, 2015, http://www.oxfordanimalethics.com/home/

58. For the Animal Legal Defense Fund's Web site, see: Animal Legal Defense Fund, accessed July 15, 2015, http://aldf.org. For the Nonhuman Rights Project's Web site, see: Nonhuman Rights Project, accessed July 15, 2015, http://www.nonhumanrightsproject.org

59. See note 30 above.

60. See, PETA, "PETA's Guide to Becoming an Activist," accessed July 15, 2015, http://www.peta.org/action/activism-guide/

Epilogue

A mosaic of pieces, we said at the start. Is there, then, a picture that begins to emerge from it? Are there any definite images surfacing from this collective reflection on theory and praxis that includes different theories of society and different explicative models? It seems so. For even a brief survey of the contributions to the volume allows one to identify some problems whose centrality is the object of an overlapping consensus, and around which revolve the majority of analyses and proposals.

The most shared aspect is the idea that the movement's focus should shift from the individual to society, from private morality to social justice—that personal choices may be important, but are hardly the stuff of which political changes are made. Hence, one finds arguments to the effect that the animal movement must learn to more fully demand change within societal structures (Aaltola); that transformative change requires institutionalization, not just individual conversion, across a range of social, political, and economic locations (Donaldson and Kymlicka); or that opposing the violence toward animals is a political problem requiring strategies to challenge the institutions (Wadiwel). Starting from this basic agreement, diverse emphases are derived from different interpretations of the social world. Thus, at the structural level, the political target can be generically capitalism as a system driven by an impersonal logic of accumulation and commodification (Cavalieri), or instead capitalism in its contemporary global, neoliberal, and biopolitical forms (Calarco), or neoliberal economy's market fundamentalism (Donaldson and Kymlicka); and, at the superstructural level, chief attention

© The Author(s) 2016
P. Cavalieri (ed.), *Philosophy and the Politics of Animal Liberation*,
DOI 10.1057/978-1-137-52120-0

can be directed to humanism (Cavalieri), or to anthropocentrism (Calarco), or to the logic of human domination (Luke), or to human supremacism (Donaldson and Kymlicka).

It is clear that the shift from individuals to society implies a change in the nature of animal advocacy. This introduces a further shared challenge, which the contributors vibrantly take up, unanimously envisaging the creation of a strong, structured mass movement which might gain visibility on the political and intellectual scene. In this case too, however, differences in the analysis of society lead to divergent opinions regarding an important aspect: should this envisaged movement actively look for alliances with progressive human causes? The positive answers may, as already noted, originate from the philosophical approach—if anthropocentrism, not speciesism, is the main target, it is natural to envisage a common fight uniting animal defenders and marginalized human groups (Calarco)—or from the political belief that, to be successful, the animal liberation movement should act as a member in good standing of the family of social justice struggles (Donaldson and Kymlicka), or can rather be tied to the idea that, in the battle against institutional oppression, it is fundamental to avoid policies that merely substitute one hierarchy for another (Wadiwel). As for the negative stances, the rationale can be found in the precedence given to the building of an autonomous social formation (Smulewicz-Zucker), or in the contention that humanism leads just the most oppressed human groups to cling to their humanity to the detriment of animals (Luke, Cavalieri).

The attitude toward possible alliances cannot but interact with the background theoretical approach in generating the responses to the last shared problem—the formulation of a global political strategy. Thus, in the liberal field, while the pro-alliance Rawlsian strand favors incremental reform through a process by which local innovations lead to context-specific coalitions able to develop an animal-friendly economy which might also actively resist neoliberalism (Donaldson and Kymlicka), the Hegelian liberal orientation focusing on an autonomous path defends an immediate pressure on the state through a large-scale formal organization which can make animal rights not only a political but also an electoral issue (Smulewicz-Zucker). And whereas from the pro-alliance radical movementist area comes the exhortation to join other radical causes in addressing injustice, so as to assemble the power to forge new modes of production capable of simultaneously overcoming both human exploitation and nonhuman commodification (Calarco), within a biopolitical frame the question of cooperation comes to the forefront via the hypothesis of a direct involvement of labor in the attempt to disrupt institutional violence through the envisioned one-day suspension of animal killing (Wadiwel). Finally, in a critico-dialectical perspective the emphasis falls on the amplification of existing contradictions which may operate as catalysts of change, creating a breach in the divide between humans as persons and citizens and nonhumans as things and property (Cavalieri).

What could one say of all this from the point of view of praxis? Certainly, the consensus on the questions to be answered is important. But aren't the explicative models and their normative implications hardly reconcilable? Isn't it too difficult to draw from this debate any, if not actual, at least potential political direction? An immediate rejoinder comes from past records: as we have seen, most movements go through a process of reflection and controversy before being able to fully develop the principles of pertinence—the common properties that lie beyond their internal diversities—on the basis of which they can construct their political identity.

On the other hand, to say that there is no conjoint immediate indication is not wholly correct. For, with variable emphasis and elaboration, in all chapters one can find the idea that a fundamental, and even accessible, area of intervention for the movement is the social realm of culture in all its diverse forms. Thus, culture in the broad anthropological sense of patterns of learned behaviors and dominant sets of habits appears in the discussion as what must be challenged in order to politically remove attitudes encouraging animal exploitation (Aaltola), as what must be altered to create a lived environment more hospitable to nonhuman beings (Donaldson and Kymlicka), or as what must be contested through the generation of counter cultures that innovate around pro-animal knowledges (Wadiwel). Culture is also targeted in its philosophical sense as the historic repository of thought within which human modes of discrimination against animals and other humans have been justified (Calarco), and in the more focused sense of an inherited body of learned and literary work that should be reconsidered and critically retold (Luke). Finally, culture is a matter of critical concern in its objectified state, that is, in the form of the places and institutions where the human-centered discourse is produced and reproduced, and, in particular, of that academy from which most contributors come, and which, as mentioned, is, at least under some circumstances, liable to undergo a sort of colonization by heterodox discourses (Cavalieri, Donaldson and Kymlicka, Smulewicz-Zucker).

If one thinks that, together with other current initiatives, this is only a propaedeutic work, essentially meant to set out some coordinates for future political elaborations, such an agreement on a multifaceted attack on the ideational, behavioral, and material aspects of the dominant culture is no small result. For it means not only that there is room for constructive discussion, but also that there is a natural area of convergence that philosophical reflection can further amplify, thus providing the movement with a possible focus around which its various souls might start to act in a coordinated way.

INDEX

© The Author(s) 2016
P. Cavalieri (ed.), *Philosophy and the Politics of Animal Liberation*,
DOI 10.1057/978-1-137-52120-0